br🧠in rules

BONUS FILM

Featuring **John Medina**

Watch the 45-minute film as an introduction to the book

Special link for paperback readers*:

www.brainrules.net/dvd

[

ABOUT THE FILM

The *Brain Rules* film is a lively tour of the 12 Brain Rules.
You will experience Medina's rare gift for making science fun,
accessible, and relevant. The film will take your understanding
of the book to the next level.

]

*Hardcover and audio books include the film on DVD

John Medina is a developmental molecular biologist and research consultant. He is an affiliate Professor of Bioengineering at the University of Washington School of Medicine. He is also the director of the Brain Center for Applied Learning Research at Seattle Pacific University. He lives in Seattle, Washington.

brain rules

12 Principles for Surviving and Thriving at Work, Home, and School

JOHN MEDINA

Pear
Press

Pear Press
P.O. Box 70525
Seattle, WA 98127-0525
U.S.A.

This book may be purchased for educational, business,
or sales promotional use. For information, please visit
www.pearpress.com.

First Pear Press trade paperback edition 2009

Edited by Tracy Cutchlow
Designed by Greg Pearson

Library of Congress Cataloging-In-Publication Data is available upon request.

ISBN-10: 0-9797777-4-7 (pbk.)
ISBN-13: 978-0-9797777-4-5 (pbk.)

ISBN-10: 0-9797777-0-4 (hc.)
ISBN-13: 978-0-9797777-0-7 (hc.)

10 9 8 7 6 5 4 3 2 1

To Joshua and Noah

Gratitude, my dear boys, for constantly reminding me
that age is not something that matters unless you are cheese.

contents

introduction

GO AHEAD AND MULTIPLY the number 8,388,628 x 2 in your head. Can you do it in a few seconds? There is a young man who can double that number *24 times* in the space of a few seconds. He gets it right every time. There is a boy who can tell you the precise time of day at any moment, even in his sleep. There is a girl who can correctly determine the exact dimensions of an object 20 feet away. There is a child who at age 6 drew such lifelike and powerful pictures, she got her own show at a gallery on Madison Avenue. Yet none of these children could be taught to tie their shoes. Indeed, none of them have an IQ greater than 50.

The brain is an amazing thing.

Your brain may not be nearly so odd, but it is no less extraordinary. Easily the most sophisticated information-transfer system on Earth, your brain is fully capable of taking the little black squiggles on this piece of bleached wood and deriving meaning from them. To accomplish this miracle, your brain sends jolts of electricity crackling through hundreds of miles of wires composed of brain cells so small that thousands of them could fit into the period at the end of this sentence. You accomplish all of this in less time than it takes you to blink. Indeed, you have just done it. What's equally incredible, given our intimate association with it, is this: Most of us have no idea how our brain works.

This has strange consequences. We try to talk on our cell phones and drive at the same time, even though it is literally impossible for our brains to multitask when it comes to paying attention. We have created high-stress office environments, even though a stressed brain is significantly less productive. Our schools are designed so that most real learning has to occur at home. This would be funny if it weren't so harmful. Blame it on the fact that brain scientists rarely have a conversation with teachers and business professionals, education majors and accountants, superintendents and CEOs. Unless you have the *Journal of Neuroscience* sitting on your coffee table, you're out of the loop.

This book is meant to get you into the loop.

12 brain rules

My goal is to introduce you to 12 things we know about how the brain works. I call these Brain Rules. For each rule, I present the science and then offer ideas for investigating how the rule might apply to our daily lives, especially at work and school. The brain is complex, and I am taking only slivers of information from each subject—not comprehensive but, I hope, accessible. The Brain Rules film, available at *www.brainrules.net/dvd*, is an integral part of the project. You might use the DVD as an introduction, and then jump between a chapter in the book and the illustrations online. A sampling of the ideas you'll encounter:

• For starters, we are not used to sitting at a desk for eight hours a day. From an evolutionary perspective, our brains developed while working out, walking as many as 12 miles a day. The brain still craves that experience, especially in sedentary populations like our own. That's why exercise boosts brain power (Brain Rule #1) in such populations. Exercisers outperform couch potatoes in long-term memory, reasoning, attention, and problem-solving tasks. I am convinced that integrating exercise into our eight hours at work or school would only be normal.

• As you no doubt have noticed if you've ever sat through a typical PowerPoint presentation, people don't pay attention to boring things (Brain Rule #4). You've got seconds to grab someone's attention and only 10 minutes to keep it. At 9 minutes and 59 seconds, something must be done to regain attention and restart the clock—something emotional and relevant. Also, the brain needs a break. That's why I use stories in this book to make many of my points.

• Ever feel tired about 3 o'clock in the afternoon? That's because your brain really wants to take a nap. You might be more productive if you did: In one study, a 26-minute nap improved NASA pilots' performance by 34 percent. And whether you get enough rest at night affects your mental agility the next day. Sleep well, think well (Brain Rule #7).

• We'll meet a man who can read two pages at the same time, one with each eye, and remember everything in the pages forever. Most of us do more forgetting than remembering, of course, and that's why we must repeat to remember (Brain Rule #5). When you understand the brain's rules for memory, you'll see why I want to destroy the notion of homework.

• We'll find out why the terrible twos only look like active rebellion but actually are a child's powerful urge to explore. Babies may not have a lot of knowledge about the world, but they know a whole lot about how to get it. We are powerful and natural explorers (Brain Rule #12), and this never leaves us, despite the artificial environments we've built for ourselves.

no prescriptions

The ideas ending the chapters of this book are not a prescription.

They are a call for real-world research. The reason springs from what I do for a living. My research expertise is the molecular basis of psychiatric disorders, but my real interest is in trying to understand the fascinating distance between a gene and a behavior. I have been a private consultant for most of my professional life, a hired gun for research projects in need of a developmental molecular biologist with such specialization. I have had the privilege of watching countless research efforts involving chromosomes and mental function.

On such journeys, I occasionally would run across articles and books that made startling claims based on "recent advances" in brain science about how to change the way we teach people and do business. And I would panic, wondering if the authors were reading some literature totally off my radar screen. I speak several dialects of brain science, and I knew nothing from those worlds capable of dictating best practices for education and business. In truth, if we ever fully understood how the human brain knew how to pick up a glass of water, it would represent a major achievement.

There was no need to panic. You can responsibly train a skeptical eye on any claim that brain research can without equivocation tell us how to become better teachers, parents, business leaders, or students. This book is a call for research simply because we don't know enough to be prescriptive. It is an attempt to vaccinate against mythologies such as the "Mozart Effect," left brain/right brain personalities, and getting your babies into Harvard by making them listen to language tapes while they are still in the womb.

back to the jungle

What we know about the brain comes from biologists who study brain tissues, experimental psychologists who study behavior, cognitive neuroscientists who study how the first relates to the second, and evolutionary biologists. Though we know precious little about how the brain works, our evolutionary history tells us this: The brain appears to be designed to solve problems related to surviving

4

in an unstable outdoor environment, and to do so in nearly constant motion. I call this the brain's performance envelope.

Each subject in this book—exercise, survival, wiring, attention, memory, sleep, stress, sense, vision, gender, and exploration—relates to this performance envelope. Motion translates to exercise. Environmental instability led to the extremely flexible way our brains are wired, allowing us to solve problems through exploration. Learning from our mistakes so we could survive in the great outdoors meant paying attention to certain things at the expense of others, and it meant creating memories in a particular way. Though we have been stuffing them into classrooms and cubicles for decades, our brains actually were built to survive in jungles and grasslands. We have not outgrown this.

I am a nice guy, but I am a grumpy scientist. For a study to appear in this book, it has to pass what some at The Boeing Company (for which I have done some consulting) call MGF: the Medina Grump Factor. That means the supporting research for each of my points must first be published in a peer-reviewed journal and then successfully replicated. Many of the studies have been replicated dozens of times. (To stay as reader-friendly as possible, extensive references are not in this book but can be found at *www.brainrules.net.*)

What do these studies show, viewed as a whole? Mostly this: If you wanted to create an education environment that was directly opposed to what the brain was good at doing, you probably would design something like a classroom. If you wanted to create a business environment that was directly opposed to what the brain was good at doing, you probably would design something like a cubicle. And if you wanted to change things, you might have to tear down both and start over.

In many ways, starting over is what this book is all about.

exercise

Rule #1
Exercise boosts brain power.

IF THE CAMERAS WEREN'T rolling and the media abuzz with live reports, it is possible nobody would have believed the following story:

A man had been handcuffed, shackled and thrown into California's Long Beach Harbor, where he was quickly fastened to a floating cable. The cable had been attached at the other end to 70 boats, bobbing up and down in the harbor, each carrying a single person. Battling strong winds and currents, the man then swam, towing all 70 boats (and passengers) behind him, traveling 1.5 miles to Queen's Way Bridge. The man, Jack La Lanne, was celebrating his birthday.

He had just turned 70 years old.

Jack La Lanne, born in 1914, has been called the godfather of the American fitness movement. He starred in one of the longest-running exercise programs produced for commercial television. A prolific inventor, La Lanne designed the first leg-extension machines, the first cable-fastened pulleys, and the first weight selectors, all now standard issue in the modern gym. He is even credited with inventing an exercise that supposedly bears his name, the Jumping Jack. La Lanne is now in his mid-90s, and even these feats are probably not the most interesting aspect of this famed bodybuilder's story.

If you ever have the chance to hear him in an interview, your biggest impression will be not the strength of his muscles but the strength of his *mind*. La Lanne is mentally alert, almost beyond reason. His sense of humor is both lightening fast and improvisatory. "I tell people I can't afford to die. It will wreck my image!" he once exclaimed to Larry King. He regularly rails at the camera: "Why am I so strong? Do you know how many calories are in butter and cheese and ice cream? Would you get your *dog* up in the morning for a cup of coffee and a doughnut?" He claims he hasn't had dessert since 1929. He is hyper-energized, opinionated, possessed with the intellectual vigor of an athlete in his 20s.

So it's hard not to ask: "Is there a relationship between exercise and mental alertness?" The answer, it turns out, is yes.

survival of the fittest

Though a great deal of our evolutionary history remains shrouded in controversy, the one fact that every paleoanthropologist on the planet accepts can be summarized in two words:

We *moved*.

A lot. When our bountiful rainforests began to shrink, collapsing the local food supply, we were forced to wander around an increasingly dry landscape looking for more trees we could scamper up to dine. As the climate got more arid, these wet botanical vending machines disappeared altogether. Instead of moving up and down complex arboreal environments in three dimensions, which required a lot of dexterity, we began walking back and forth across arid savannahs in two dimensions, which required a lot of stamina.

"About 10 to 20 kilometers a day with men," says famed anthropologist Richard Wrangham, "and about half that for women."

That's the amount of ground scientists estimate we covered on a *daily* basis back then—up to 12 miles a day. That means our fancy brains developed not while we were lounging around but while we were working out.

The first real marathon runner of our species was a vicious predator known as *Homo erectus*. As soon as the *Homo erectus* family evolved, about 2 million years ago, he started moving out of town. Our direct ancestors, *Homo sapiens,* rapidly did the same thing, starting in Africa 100,000 years ago and reaching Argentina by 12,000 years ago. Some researchers suggest that we were extending our ranges by an unheard-of 25 miles per year.

This is an impressive feat, considering the nature of the world our ancestors inhabited. They were crossing rivers and deserts, jungles and mountain ranges, all without the aid of maps and mostly without tools. They eventually made ocean-going boats without the benefit of wheels or metallurgy, and then traveling up and down the Pacific with only the crudest navigational skills. Our ancestors constantly were encountering new food sources, new predators, new physical dangers. Along the road they routinely suffered injuries, experienced strange illnesses, and delivered and nurtured children, all without the benefit of textbooks or modern medicine.

Given our relative wimpiness in the animal kingdom (we don't even have enough body hair to survive a mildly chilly night), what these data tell us is that we grew up in top physical shape, or we didn't grow up at all. And they also tell us the human brain became the most powerful in the world under conditions where motion was a constant presence.

If our unique cognitive skills were forged in the furnace of physical activity, is it possible that physical activity still influences our cognitive skills? Are the cognitive abilities of someone in good physical condition different from those of someone in poor physical condition? And what if someone in poor physical condition were whipped into shape? Those are scientifically testable questions. The answers are directly related to why Jack La Lanne can still crack jokes about eating dessert. *In his nineties.*

will you age like jim or like frank?

We discovered the beneficial effects of exercise on the brain by looking at aging populations. This was brought home to me by an anonymous man named Jim and a famous man named Frank. I met them both while I was watching television. A documentary on American nursing homes showed people in wheelchairs, many in their mid- to late 80s, lining the halls of a dimly lit facility, just sitting around, seemingly waiting to die. One was named Jim. His eyes seemed vacant, lonely, friendless. He could cry at the drop of a hat but otherwise spent the last years of his life mostly staring off into space. I switched channels. I stumbled upon a very young-looking Mike Wallace. The journalist was busy interviewing architect Frank Lloyd Wright, at the time in his late 80s. I was about to hear a most riveting interview.

"When I walk into St. Patrick's Cathedral ... here in New York City, I am enveloped in a feeling of reverence," said Wallace, tapping his cigarette. The old man eyed Wallace. "Sure it isn't an inferiority complex?"

"Just because the building is big and I'm small, you mean?"

"Yes."

"I think not."

"I hope not."

"You feel nothing when you go into St. Patrick's?"

"Regret," Wright said without a moment's pause, "because it isn't the thing that really represents the spirit of independence and the sovereignty of the individual which I feel should be represented in our edifices devoted to culture."

I was dumbfounded by the dexterity of Wright's response. In four sentences, one could detect the clarity of his mind, his unshakable vision, his willingness to think out of the box. The rest of his interview was just as compelling, as was the rest of Wright's life. He completed the designs for the Guggenheim Museum, his last work, in 1957, when he was 90 years old.

But I also was dumbfounded by something else. As I contemplated Wright's answers, I remembered Jim from the nursing home. *He was the same age as Wright.* In fact, most of the residents were. I suddenly was beholding two types of aging. Jim and Frank lived in roughly the same period of time. But one mind had almost completely withered, while the other remained as incandescent as a light bulb. What was the difference in the aging process between men like Jim and the famous architect? This question has bugged the research community for a long time. Investigators have known for years that some people age with energy and pizazz, living productive lives well into their 80s and 90s. Others appear to become battered and broken by the process, and often they don't survive their 70s. Attempts to explain these differences led to many important discoveries, which I have grouped as answers to six questions.

1) Is there one factor that predicts how well you will age?

It was never an easy question for researchers to answer. They found many variables, from nature to nurture, that contributed to someone's ability to age gracefully. That's why the scientific community met with both applause and suspicion a group of researchers who uncovered a powerful environmental influence. In a result that probably produced a smile on Jack La Lanne's face, one of the greatest predictors of successful aging was the presence or absence of a sedentary lifestyle. Put simply, if you are a couch potato, you are more likely to age like Jim, if you make it to your 80s at all. If you have an active lifestyle, you are more likely to age like Frank Lloyd Wright and much more likely to make it to your 90s.

The chief reason for the difference seemed to be that exercise improved cardiovascular fitness, which in turn reduced the risk for diseases such as heart attacks and stroke. But researchers wondered why the people who were aging "successfully" also seemed to be more mentally alert. This led to the obvious second question:

2) Were they?

Just about every mental test possible was tried. No matter how it was measured, the answer was consistently yes: A lifetime of exercise can result in a sometimes astonishing elevation in cognitive performance, compared with those who are sedentary. Exercisers outperform couch potatoes in tests that measure long-term memory, reasoning, attention, problem-solving, even so-called fluid-intelligence tasks. These tasks test the ability to reason quickly and think abstractly, improvising off previously learned material in order to solve a new problem. Essentially, exercise improves a whole host of abilities prized in the classroom and at work.

Not every weapon in the cognitive arsenal is improved by exercise. Short-term memory skills, for example, and certain types of reaction times appear to be unrelated to physical activity. And, while nearly everybody shows some improvement, the degree of benefit varies quite a bit among individuals. Most important, these data, strong as they were, showed only an association, not a cause. To show the direct link, a more intrusive set of experiments had to be done. Researchers had to ask:

3) Can you turn Jim into Frank?

The experiments were reminiscent of a makeover show. Researchers found a group of couch potatoes, measured their brain power, exercised them for a period of time, and re-examined their brain power. They consistently found that when couch potatoes are enrolled in an aerobic exercise program, all kinds of mental abilities begin to come back online. Positive results were observed after as little as four months of activity. It was the same story with school-age children. In one recent study, children jogged for 30 minutes two or three times a week. After 12 weeks, their cognitive performance had improved significantly compared with pre-jogging levels. When the exercise program was withdrawn, the scores plummeted back to their pre-experiment levels. Scientists had found a direct link. Within limits, it does appear that exercise can turn Jim into Frank, or at least turn Jim into a sharper version of himself.

As the effects of exercise on cognition became increasingly obvious, scientists began fine-tuning their questions. One of the biggest—certainly one dearest to the couch-potato cohort—was: What type of exercise must you do, and how much of it must be done to get the benefit? I have both good news and bad news.

4) What's the bad news?

Astonishingly, after years of investigation in aging populations, the answer to the question of how much is *not much*. If all you do is walk several times a week, your brain will benefit. Even couch potatoes who fidget show increased benefit over those who do not fidget. The body seems to be clamoring to get back to its hyperactive Serengeti roots. Any nod toward this history, be it ever so small, is met with a cognitive war whoop. In the laboratory, the gold standard appears to be aerobic exercise, 30 minutes at a clip, two or three times a week. Add a strengthening regimen and you get even more cognitive benefit.

Of course, individual results vary, and no one should embark on a rigorous program without consulting a physician. Too much exercise and exhaustion can hurt cognition. The data merely point to the fact that one should embark. Exercise, as millions of years traipsing around the backwoods tell us, is good for the brain. Just how good took everyone by surprise, as they answered the next question.

5) Can exercise treat brain disorders?

Given the robust effect of exercise on typical cognitive performance, researchers wanted to know if it could be used to treat atypical performance. What about diseases such as age-related dementia and its more thoroughly investigated cousin, Alzheimer's disease? What about affective disorders such as depression? Researchers looked at both prevention and intervention. With experiments reproduced all over the world, enrolling thousands of people, often studied for decades, the results are clear. Your lifetime risk for general dementia is literally cut in half if you participate in leisure-time physical activity. Aerobic exercise seems to be the key. With Alzheimer's, the effect is even greater: Such exercise lowers your odds of getting the disease by more than 60 percent.

How much exercise? Once again, a little goes a long way. The researchers showed you have to participate in some form of exercise just twice a week to get the benefit. Bump it up to a 20-minute walk each day, and you can cut your risk of having a stroke—one of the leading causes of mental disability in the elderly—by 57 percent.

The man most responsible for stimulating this line of inquiry did not start his career wanting to be a scientist. He wanted to be an athletics coach. His name is Dr. Steven Blair, and he looks uncannily like Jason Alexander, the actor who portrayed George Costanza on the old TV sitcom *Seinfeld*. Blair's coach in high school, Gene Bissell, once forfeited a football game after discovering that an official had missed a call. Even though the league office balked, Bissell insisted that his team be declared the loser, and the young Steven never forgot the incident. Blair writes that this devotion to truth inspired his undying admiration for rigorous, no-nonsense, statistical analysis of the epidemiological work in which he eventually embarked. His seminal paper on fitness and mortality stands as a landmark example of how to do work with integrity in this field. The rigor of his findings inspired other investigators. What about using exercise not only as prevention, they asked, but as intervention, to treat mental disorders such as depression and anxiety?

That turned out to be a good line of questioning. A growing body of work now suggests that physical activity can powerfully affect the course of both diseases. We think it's because exercise regulates the release of the three neurotransmitters most commonly associated with the maintenance of mental health: serotonin, dopamine, and norepinephrine. Although exercise cannot substitute for psychiatric treatment, the role of exercise on mood is so pronounced that many psychiatrists have begun adding a regimen of physical activity to the normal course of therapy. But in one experiment with depressed individuals, rigorous exercise was actually substituted for antidepressant medication. Even when compared against medicated controls, the treatment outcomes were astonishingly successful. For both depression and anxiety, exercise is beneficial immediately and over the long term. It is equally effective for men and women, and the longer the program is deployed, the greater the effect becomes. It is especially helpful for severe cases and for older people.

Most of the data we have been discussing concern elderly populations. Which leads to the question:

6) Are the cognitive blessings of exercise only for the elderly?

As you ratchet down the age chart, the effects of exercise on cognition become less clear. The biggest reason for this is that so few studies have been done. Only recently has the grumpy scientific eye begun to cast its gaze on younger populations. One of the best efforts enrolled more than 10,000 British civil servants between the ages of 35 and 55, examining exercise habits and grading them as low, medium, or high. Those with low levels of physical activity were more likely to have poor cognitive performance. Fluid intelligence, the type that requires improvisatory problem-solving skills, was particularly hurt by a sedentary lifestyle. Studies done in other countries have confirmed the finding.

If only a small number of studies have been done in middle-age populations, the number of studies saying anything about exercise and children is downright microscopic. Though much more work needs to be done, the data point in a familiar direction, though perhaps for different reasons.

To talk about some of these differences, I would like to introduce you to Dr. Antronette Yancey. At 6 foot 2, Yancey is a towering, beautiful presence, a former professional model, now a physician-scientist with a deep love for children and a broad smile to buttress the attitude. She is a killer basketball player, a published poet, and one of the few professional scientists who also makes performance art. With this constellation of talents, she is a natural to study the effects of physical activity on developing minds. And she has found what everybody else has found: Exercise improves children. Physically fit children identify visual stimuli much faster than sedentary ones. They appear to concentrate better. Brain-activation studies show that children and adolescents who are fit allocate more cognitive resources to a task and do so for longer periods of time.

"Kids pay better attention to their subjects when they've been active," Yancey says. "Kids are less likely to be disruptive in terms of their classroom behavior when they're active. Kids feel better about themselves, have higher self-esteem, less depression, less anxiety. All of those things can impair academic performance and attentiveness."

Of course, there are many ingredients to the recipe of academic performance. Finding out which components are the most important—especially if you want improvement—is difficult enough. Finding out whether exercise is one of those choice ingredients is even tougher. But these preliminary findings show that we have every reason to be optimistic about the long-term outcomes.

an exercise in road-building

Why exercise works so well in the brain, at a molecular level, can be explained by competitive food eaters—or, less charitably, professional pigs. There is an international association representing people who time themselves on how much they can eat at a given event. The association is called the International Federation of Competitive Eating, and its crest proudly displays the slogan (I am not making this up) *In Voro Veritas*—literally, "In Gorging, Truth."

Like any sporting organization, competitive food eaters have their heroes. The reigning gluttony god is Takeru "Tsunami" Kobayashi. He is the recipient of many eating awards, including the vegetarian dumpling competition (83 dumplings downed in 8 minutes), the roasted pork bun competition (100 in 12 minutes), and the hamburger competition (97 in 8 minutes). Kobayashi also is a world champion hot-dog eater. One of his few losses was to a 1,089-pound Kodiak bear. In a 2003 Fox televised special called *Man vs. Beast,* the mighty Kobayashi consumed only 31 bunless dogs compared with the ursine's 50, all in about 2½ minutes. Kobayashi lost his hot-dog crown in 2007 to Joey Chestnut, who ate 66 hot dogs in 12 minutes (the Tsunami could manage only 63).

But my point isn't about speed. It's about what happens to all of those hot dogs after they slide down the Tsunami's throat. As with any of us, his body uses its teeth and acid and wormy intestines to tear the food apart and, if need be, reconfigure it.

This is done for more or less a single reason: to turn foodstuffs into glucose, a type of sugar that is one of the body's favorite energy resources. Glucose and other metabolic products are absorbed into the bloodstream via the small intestines. The nutrients travel to all parts of the body, where they are deposited into cells, which make up the body's various tissues. The cells seize the sweet stuff like sharks in a feeding frenzy. Cellular chemicals greedily tear apart the molecular structure of glucose to extract its sugary energy. This energy extraction is so violent that atoms are literally ripped asunder in the process.

As in any manufacturing process, such fierce activity generates a fair amount of toxic waste. In the case of food, this waste consists of a nasty pile of excess electrons shredded from the atoms in the glucose molecules. Left alone, these electrons slam into other molecules within the cell, transforming them into some of the most toxic substances known to humankind. They are called free radicals. If not quickly corralled, they will wreck havoc on the innards of a cell and, cumulatively, on the rest of the body. These electrons are fully capable, for example, of causing mutations in your very DNA.

The reason you don't die of electron overdose is that the atmosphere is full of breathable oxygen. The main function of oxygen is to act like an efficient electron-absorbing sponge. At the same time the blood is delivering foodstuffs to your tissues, it is also carrying these oxygen sponges. Any excess electrons are absorbed by the oxygen and, after a bit of molecular alchemy, are transformed into equally hazardous—but now fully transportable—carbon dioxide. The blood is carried back to your lungs, where the carbon dioxide leaves the blood and you breathe it out. So, whether you are a competitive eater or a typical one, the oxygen-rich air you inhale keeps the food you eat from killing you.

Getting food into tissues and getting toxic electrons out obviously are matters of access. That's why blood has to be everywhere inside you. Serving as both wait staff and haz-mat team, any tissue without enough blood supply is going to starve to death—your brain included. That's important because the brain's appetite for energy is enormous. The brain represents only about 2 percent of most people's body weight, yet it accounts for about 20 percent of the body's total energy usage—about 10 times more than would be expected. When the brain is fully working, it uses more energy per unit of tissue weight than a fully exercising quadricep. In fact, the human brain cannot simultaneously activate more than 2 percent of its neurons at any one time. More than this, and the glucose supply becomes so quickly exhausted that you will faint.

If it sounds to you like the brain needs a lot of glucose—and generates a lot of toxic waste—you are right on the money. This means the brain also needs lots of oxygen-soaked blood. How much food and waste can the brain generate in just a few minutes? Consider the following statistics. The three requirements for human life are food, drink, and fresh air. But their effects on survival have very different timelines. You can live for 30 days or so without food, and you can go for a week or so without drinking water. Your brain, however, is so active that it cannot go without oxygen for more than 5 minutes without risking serious and permanent damage. Toxic electrons over-accumulate because the blood can't deliver enough oxygen sponges. Even in a healthy brain, the blood's delivery system can be improved. That's where exercise comes in. It reminds me of a seemingly mundane little insight that literally changed the history of the world.

The man with the insight was named John Loudon McAdam. McAdam, a Scottish engineer living in England in the early 1800s, noticed the difficulty people had trying to move goods and supplies over hole-filled, often muddy, frequently impassable dirt roads. He got the splendid idea of raising the level of the road using layers of rock and gravel. This immediately made the roads more stable, less muddy, and less flood-prone. As county after county adopted his process, now called macadamization, an astonishing after-effect occurred. People instantly got more dependable access to one another's goods and services. Offshoots from the main roads sprang up, and pretty soon entire countrysides had access to far-flung points using stable arteries of transportation. Trade grew. People got richer. By changing the way things moved, McAdam changed the way we lived. What does this have to do with exercise? McAdam's central notion wasn't to improve goods and services, but to improve *access* to goods and services. You can do the same for your brain by increasing the roads in your body, namely your blood vessels, through exercise. Exercise does not provide the oxygen and the food. It provides your body greater *access* to the oxygen and the food. How this works is easy to understand.

When you exercise, you increase blood flow across the tissues of your body. This is because exercise stimulates the blood vessels to create a powerful, flow-regulating molecule called nitric oxide. As the flow improves, the body makes new blood vessels, which penetrate deeper and deeper into the tissues of the body. This allows more access to the bloodstream's goods and services, which include food distribution and waste disposal. The more you exercise, the more tissues you can feed and the more toxic waste you can remove. This happens all over the body. That's why exercise improves the performance of most human functions. You stabilize existing transportation structures and add new ones, just like McAdam's roads. All of a sudden, you are becoming healthier.

The same happens in the human brain. Imaging studies have shown that exercise literally increases blood volume in a region of the brain called the dentate gyrus. That's a big deal. The dentate gyrus is a vital constituent of the hippocampus, a region deeply involved in memory formation. This blood-flow increase, which may be the result of new capillaries, allows more brain cells greater access to the blood's food and haz-mat teams.

Another brain-specific effect of exercise recently has become clear, one that isn't reminiscent of roads so much as of fertilizer. At the molecular level, early studies indicate that exercise also stimulates one of the brain's most powerful growth factors, BDNF. That stands for Brain Derived Neurotrophic Factor, and it aids in the development of healthy tissue. BDNF exerts a fertilizer-like growth effect on certain neurons in the brain. The protein keeps existing neurons young and healthy, rendering them much more willing to connect with one another. It also encourages neurogenesis, the formation of new cells in the brain. The cells most sensitive to this are in the hippocampus, inside the very regions deeply involved in human cognition. Exercise increases the level of usable BDNF inside those cells. The more you exercise, the more fertilizer you create—at least, if you are a laboratory animal. There are now suggestions that the same mechanism also occurs in humans.

we can make a comeback

All of the evidence points in one direction: Physical activity is cognitive candy. We can make a species-wide athletic comeback.

All we have to do is move. When people think of great comebacks, athletes such as Lance Armstrong or Paul Hamm usually come to mind. One of the greatest comebacks of all time, however, occurred before both of these athletes were born. It happened in 1949 to the legendary golfer Ben Hogan.

Prickly to the point of being obnoxious (he once quipped of a competitor, "If we could have just screwed another head on his shoulders, he would have been the greatest golfer who ever lived"), Hogan's gruff demeanor underscored a fierce determination. He won the PGA championship in 1946 and in 1948, the year in which he was also named PGA Player of the Year. That all ended abruptly. On a foggy night in the Texas winter of 1949, Hogan and his wife were hit head-on by a bus. Hogan fractured every bone that could matter to a golfer: collar bone, pelvis, ankle, rib. He was left with life-threatening blood clots. The doctors said he might never walk again, let alone play golf. Hogan ignored their prognostications. A year after the accident, he climbed back onto the green and won the U.S. Open. Three years later, he played one of the most successful single seasons in professional golf. He won five of the six tournaments he entered, including the first three major championships of the year (a feat now known as the Hogan Slam). Reflecting on one of the greatest comebacks in sports history, he said in his typically spicy manner, "People have always been telling me what I can't do." He retired in 1971.

When I reflect on the effects of exercise on cognition and the things we might try to recapture its benefits, I am reminded of such comebacks. Civilization, while giving us such seemingly forward advances as modern medicine and spatulas, also has had a nasty side effect. It gave us more opportunities to sit on our butts. Whether learning or working, we gradually quit exercising the way our ancestors did. The result is like a traffic wreck.

Recall that our evolutionary ancestors were used to walking up to 12 miles *per day.* This means that our brains were supported for most of our evolutionary history by Olympic-caliber bodies. We were not used to sitting in a classroom for 8 hours at a stretch. We were not used to sitting in a cubicle for 8 hours at a stretch. If we sat around the Serengeti for 8 hours—heck, for 8 *minutes*—we were usually somebody's lunch. We haven't had millions of years to adapt to our sedentary lifestyle. That means we need a comeback. Removing ourselves from such inactivity is the first step. I am convinced that integrating exercise into those 8 hours at work or school will not make us *smarter*. It will only make us *normal*.

ideas

There is no question we are in an epidemic of fatness, a point I will not belabor here. The benefits of exercise seem nearly endless because its impact is systemwide, affecting most physiological systems. Exercise makes your muscles and bones stronger, for example, and improves your strength and balance. It helps regulate your appetite, changes your blood lipid profile, reduces your risk for more than a dozen types of cancer, improves the immune system, and buffers against the toxic effects of stress (see Chapter 8). By enriching your cardiovascular system, exercise decreases your risk for heart disease, stroke, and diabetes. When combined with the intellectual benefits exercise appears to offer, we have in our hands as close to a magic bullet for improving human health as exists in modern medicine. There must be ways to harness the effects of exercise in the practical worlds of education and business.

Recess twice a day

Because of the increased reliance on test scores for school survival, many districts across the nation are getting rid of physical education and recess. Given the powerful cognitive effects of physical activity, this makes no sense. Yancey, the model-turned-physician/scientist/basketball player, describes a real-world test:

"They took time away from academic subjects for physical

education … and found that, across the board, [physical education] did not hurt the kids' performance on the academic tests. … [When] trained teachers provided the physical education, the children actually did better on language, reading and the basic battery of tests."

Cutting off physical exercise—the very activity most likely to promote cognitive performance—to do better on a test score is like trying to gain weight by starving yourself. What if a school district inserted exercise into the normal curriculum on a regular basis, even twice a day? After all of the children had been medically evaluated, they'd spend 20 to 30 minutes each morning on formal aerobic exercise; in the afternoon, 20 to 30 minutes on strengthening exercises. Most populations studied see a benefit if this is done only two or three times a week. If it worked, there would be many ramifications. It might even reintroduce the notion of school uniforms. Of what would the new apparel consist? Simply gym clothes, worn all day long.

Treadmills in classrooms and cubicles

Remember the experiment showing that when children aerobically exercised, their brains worked better, and when the exercise was withdrawn, the cognitive gain soon plummeted? These results suggested to the researchers that the level of fitness was not as important as a steady increase in the oxygen supply to the brain (otherwise the improved mental sharpness would not have fallen off so rapidly). So they did another experiment. They found that supplemental oxygen administered to young healthy adults without exercise gave a similar cognitive improvement.

This suggests an interesting idea to try in a classroom (don't worry, it doesn't involve oxygen doping to get a grade boost). What if, during a lesson, the children were not sitting at desks but walking on treadmills? Students might listen to a math lecture while walking 1 to 2 miles per hour, or study English on treadmills fashioned to

25

accommodate a desktop. Treadmills in the classroom might harness the valuable advantages of increasing the oxygen supply naturally and at the same time harvest all the other advantages of regular exercise. Would such a thing, deployed over a school year, change academic performance? Until brain scientists and education scientists get together to show real-world benefit, the answer is: Nobody knows.

The same idea could apply at work, with companies installing treadmills and encouraging morning and afternoon breaks for exercise. Board meetings might be conducted while people walked 2 miles per hour. Would that improve problem-solving? Would it alter retention rates or change creativity the same way it does in the laboratory?

The idea of integrating exercise into the workday may sound foreign, but it's not difficult. I put a treadmill in my own office, and I now take regular breaks filled not with coffee but with exercise. I even constructed a small structure upon which my laptop fits so I can write email while I exercise. At first, it was difficult to adapt to such a strange hybrid activity. It took a whopping 15 minutes to become fully functional typing on my laptop while walking 1.8 miles per hour.

I'm not the only one thinking along these lines. Boeing, for example, is starting to take exercise seriously in its leadership-training programs. Problem-solving teams used to work late into the night; now, all work has to be completed during the day so there's time for exercise and sleep. More teams are hitting all of their performance targets. Boeing's vice president of leadership has put a treadmill in her office as well, and she reports that the exercise clears her mind and helps her focus. Company leaders are now thinking about how to integrate exercise into working hours.

There are two compelling business reasons for such radical ideas. Business leaders already know that if employees exercised regularly, it would reduce health-care costs. And there's no question that cutting in half someone's lifetime risk of a debilitating stroke or Alzheimer's disease is a wonderfully humanitarian thing to do. But exercise also could boost the collective brain power of an organization. Fit employees are capable of mobilizing their God-given IQs better than sedentary employees. For companies whose competitiveness rests on creative intellectual horsepower, such mobilization could mean a strategic advantage. In the laboratory, regular exercise improves—sometimes dramatically so—problem-solving abilities, fluid intelligence, even memory. Would it do so in business settings? What types of exercise need to be done, and how often? That's worth investigating.

Summary

Rule #1
Exercise boosts brain power.

* Our brains were built for walking—12 miles a day!

* To improve your thinking skills, *move*.

* Exercise gets blood to your brain, bringing it glucose for energy and oxygen to soak up the toxic electrons that are left over. It also stimulates the protein that keeps neurons connecting.

* Aerobic exercise just twice a week halves your risk of general dementia. It cuts your risk of Alzheimer's by 60 percent.

Get illustrations, audio, video, and more
at www.brainrules.net

survival

Rule #2
The human brain evolved, too.

WHEN HE WAS 4, my son Noah picked up a stick in our backyard and showed it to me. "Nice stick you have there, young fellow," I said. He replied earnestly, "That's not a stick. That's a sword! Stick 'em up!" And I raised my hands to the air. We both laughed. The reason I remember this short exchange is that as I went back into the house, I realized my son had just displayed virtually every unique thinking ability a human possesses—one that took several million years to manufacture. And he did so in less than two seconds.

Heavy stuff for a 4-year-old. Other animals have powerful cognitive abilities, too, and yet there is something qualitatively different about the way humans think about things. The journey that brought us from the trees to the savannah gave us some structural elements shared by no other creature—and unique ways of using the elements we do have in common. How and why did our brains evolve this way?

Recall the performance envelope: The brain appears to be

designed to (1) solve problems (2) related to surviving (3) in an unstable outdoor environment, and (4) to do so in nearly constant motion. The brain adapted this way simply as a survival strategy, to help us live long enough to pass our genes on to the next generation. That's right: It all comes down to sex. Ecosystems are harsh, crushing life as easily as supporting it. Scientists estimate 99.99% of all species that have ever lived are extinct today. Our bodies, brains included, latched on to any genetic adaptation that helped us survive. This not only sets the stage for all of the Brain Rules, it explains how we came to conquer the world.

There are two ways to beat the cruelty of the environment: You can become stronger or you can become smarter. We chose the latter. It seems most improbable that such a physically weak species could take over the planet not by adding muscles to our skeletons but by adding neurons to our brains. But we did, and scientists have spent a great deal of effort trying to figure out how. Judy DeLoache has studied this question extensively. She became a well-respected researcher in an era when women were actively discouraged from studying investigative science, and she is still going strong at the University of Virginia. Her research focus, given her braininess? Appropriately, it is human braininess itself. She is especially interested in how human cognition can be distinguished from the way other animals think about their respective worlds.

One of her major contributions was to identify the human trait that really does separate us from the gorillas: the ability to use symbolic reasoning. That's what my son was doing when he brandished his stick sword. When we see a five-sided geometric shape, we're not stuck perceiving it as a pentagon. We can just as easily perceive the U.S. military headquarters. Or a Chrysler minivan. Our brain can behold a symbolic object as real all by itself and yet, simultaneously, also representing something else. Maybe some*things* else. DeLoache calls it Dual Representational Theory. Stated formally, it describes our ability to attribute characteristics and meanings to things that don't actually possess them. Stated informally, we can make things up that aren't there. We are human because we can fantasize.

Draw a vertical line in your hand. Does it have to stay a vertical line? Not if you know how to impute a characteristic onto something it does not intrinsically possess. Go ahead and put a horizontal line under it. Now you have the number 1. Put a dot on the top of it. Now you have the letter "i." The line doesn't have to mean a line. The line can mean anything you darn well think it should mean. The meaning can become anchored to a symbol simply because it is not forced to become anchored to anything else. The only thing you have to do is get everybody to agree on what a symbol should mean.

We are so good at dual representation, we combine symbols to derive layers of meaning. It gives us the capacity for language, and for writing down that language. It gives us the capacity to reason mathematically. It gives us the capacity for art. Combinations of circles and squares become geometry and Cubist paintings. Combinations of dots and squiggles become music and poetry. There is an unbroken intellectual line between symbolic reasoning and the ability to create culture. And no other creature is capable of doing it.

This ability isn't fully formed at birth. DeLoache was able to show this in a powerful way. In DeLoache's laboratory, a little girl plays with a dollhouse. Next door is an identical room, but life-size. DeLoache takes a small plastic dog and puts it under the dollhouse couch, then encourages the child to go into the "big" living room next door and find a "big" version of the dog. What does the little girl do? If she is 36 months of age, DeLoache found, she immediately goes to the big room, looks under the couch, and finds the big dog. But if the child is 30 months old, she has no idea where to look. She cannot reason symbolically and cannot connect the little room with the big room. Exhaustive studies show that symbolic reasoning, this all-important human trait, takes almost three years of experience to become fully operational. We don't appear to do much to distinguish ourselves from apes before we are out of the terrible twos.

a handy trait

Symbolic reasoning turned out to be a versatile gadget. Our evolutionary ancestors didn't have to keep falling into the same quicksand pit if they could tell others about it; even better if they learned to put up warning signs. With words and language, we could extract a great deal of knowledge about our living situation without always having to experience its harsh lessons directly. So it makes sense that once our brains developed symbolic reasoning, we kept it. The brain is a biological tissue; it follows the rules of biology. And there's no bigger rule in biology than evolution through natural selection: Whoever gets the food survives; whoever survives gets to have sex; and whoever has sex gets to pass his traits on to the next generation. But what stages did we go through to reach that point? How can we trace the rise of our plump, 3-pound intellects?

You might remember those old posters showing the development of humankind as a series of linear and increasingly sophisticated creatures. I have an old one in my office. The first drawing shows a chimpanzee; the final drawing shows a 1970s businessman. In between are strangely blended variations of creatures with names like Peking man and Australopithecus. There are two problems with this drawing. First, almost everything about it is wrong. Second, nobody really knows how to fix the errors. One of the biggest reasons for our lack of knowledge is that so little hard evidence exists. Most of the fossilized bones that have been collected from our ancestors could fit into your garage, with enough room left over for your bicycle and lawn mower. DNA evidence has been helpful, and there is strong evidence that we came from Africa somewhere between 7 million and 10 million years ago. Virtually everything else is disputed by some cranky professional somewhere.

Understanding our intellectual progress has been just as difficult.

Most of it has been charted by using the best available evidence: tool-making. That's not necessarily the most accurate way; even if it were, the record is not very impressive. For the first few million years, we mostly just grabbed rocks and smashed them into things. Scientists, perhaps trying to salvage some of our dignity, called these stones hand axes. A million years later, our progress still was not very impressive. We still grabbed "hand axes," but we began to smash them into other rocks, making them more pointed. Now we had sharper rocks.

It wasn't much, but it was enough to begin untethering ourselves from our East African womb, and indeed any other ecological niche. Things got more impressive, from creating fire to cooking food. Eventually, we migrated out of Africa in successive waves, our first direct *Homo sapien* ancestors making the journey as little as 100,000 years ago. Then, 40,000 years ago, something almost unbelievable happened. They appeared suddenly to have taken up painting and sculpture, creating fine art and jewelry. No one knows why the changes were so abrupt, but they were profound. Thirty-seven thousand years later, we were making pyramids. Five thousand years after that, rocket fuel.

What happened to start us on our journey? Could the growth spurt be explained by the onset of dual-representation ability? The answer is fraught with controversy, but the simplest explanation is by far the clearest. It seems our great achievements mostly had to do with a nasty change in the weather.

new rules for survival

Most of human prehistory occurred in climates like the jungles of South America: steamy, humid, and in dire need of air conditioning. It was comfortably predictable. Then the climate changed. Scientists estimate that there have been no fewer than 17 Ice Ages in the past 40 million years. Only in a few places, such as the Amazonian and African rainforests, does anything like our original, sultry, millions-of-years-old climate survive. Ice cores taken from Greenland show that the climate staggers from being unbearably hot to being sadistically cold. As little as 100,000 years ago, you could be born in a nearly arctic environment but then, mere decades later, be taking off your loincloth to catch the golden rays of the grassland sun.

Such instability was bound to have a powerful effect on any creature forced to endure it. Most could not. The rules for survival were changing, and a new class of creatures would start to fill the vacuum created as more and more of their roommates died out. That was the crisis our ancestors faced as the tropics of Northern and Eastern Africa turned to dry, dusty plains—not immediately, but inexorably—beginning maybe 10 million years ago. Some researchers blame it on the Himalayas, which had reached such heights as to disturb global atmospheric currents. Others blame the sudden appearance of the Isthmus of Panama, which changed the mixing of the Pacific and Atlantic ocean currents and disturbed global weather patterns, as El Niños do today.

Whatever the reason, the changes were powerful enough to disrupt the weather all over the world, including in our African birthplace. But not too powerful, or too subtle—a phenomenon called the Goldilocks Effect. If the changes had been too sudden, the climatic violence would have killed our ancestors outright, and I wouldn't be writing this book for you today. If the changes had been too slow, there may have been no reason to develop our talent for symbolism and, once again, no book. Instead, like Goldilocks and the third bowl of porridge, the conditions were just right. The change was enough to shake us out of our comfortable trees, but it wasn't enough to kill us when we landed.

Landing was only the beginning of the hard work, however. We quickly discovered that our new digs were already occupied. The locals had co-opted the food sources, and most of them were stronger and faster than we were. Faced with grasslands rather than trees, we rudely were introduced to the idea of "flat." It is disconcerting to think that we started our evolutionary journey on an unfamiliar horizontal plane with the words "Eat me, I'm prey" taped to the back of our evolutionary butts.

jazzin' on a riff

You might suspect that the odds against our survival were great. You would be right. The founding population of our direct ancestors is not thought to have been much larger than 2,000 individuals; some think the group was as small as a few hundred. How, then, did we go from such a wobbly, fragile minority population to a staggering tide of humanity 7 billion strong and growing? There is only one way, according to Richard Potts, director of the Human Origins Program at the Smithsonian's National Museum of Natural History. You give up on stability. You don't try to beat back the changes. You begin not to care about consistency within a given habitat, because such consistency isn't an option. You adapt to variation itself.

It was a brilliant strategy. Instead of learning how to survive in just one or two ecological niches, we took on the entire globe. Those unable to rapidly solve new problems or learn from mistakes didn't survive long enough to pass on their genes. The net effect of this evolution was that we didn't become stronger; we became smarter. We learned to grow our fangs not in the mouth but in the head. This turned out to be a pretty savvy strategy. We went on to conquer the small rift valleys in Eastern Africa. Then we took over the world.

Potts calls his notion Variability Selection Theory, and it attempts to explain why our ancestors became increasingly allergic to inflexibility and stupidity. Little in the fossil record is clear about the exact progression—another reason for bitter controversy—but all researchers must contend with two issues. One is bipedalism; the other has to do with our increasingly big heads.

Variability Selection Theory predicts some fairly simple things about human learning. It predicts there will be interactions between two powerful features of the brain: a database in which to store a fund of knowledge, and the ability to improvise off of that database. One allows us to know when we've made mistakes. The other allows us to learn from them. Both give us the ability to add new information under rapidly changing conditions. Both may be relevant to the way we design classrooms and cubicles.

Any learning environment that deals with only the database instincts or only the improvisatory instincts ignores one half of our ability. It is doomed to fail. It makes me think of jazz guitarists: They're not going to make it if they know a lot about music theory but don't know how to jam in a live concert. Some schools and workplaces emphasize a stable, rote-learned database. They ignore the improvisatory instincts drilled into us for millions of years. Creativity suffers. Others emphasize creative usage of a database, without installing a fund of knowledge in the first place. They ignore our need to obtain a deep understanding of a subject, which includes memorizing and storing a richly structured database. You get people who are great improvisers but don't have depth of knowledge. You may know someone like this where you work. They may look like jazz musicians and have the appearance of jamming, but in the end they know nothing. They're playing intellectual air guitar.

standing tall

Variability Selection Theory allows a context for dual representation, but it hardly gets us to the ideas of Judy DeLoache and our unique ability to invent calculus and write romance novels. After all, many animals create a database of knowledge, and many of them make tools, which they even use creatively. Still, it is not as if chimpanzees write symphonies badly and we write them well. Chimps can't write them at all, and we can write ones that make people spend their life savings on subscriptions to the New York Philharmonic. There must have been something else in our evolutionary history that made human thinking unique.

One of the random genetic mutations that gave us an adaptive advantage involved learning to walk upright. The trees were gone or going, so we had to deal with something new in our experience: walking increasingly long distances between food sources. That eventually involved the specialized use of our two legs. Bipedalism was an excellent solution to a vanishing rainforest. But it was also a major change. At the very least, it meant refashioning the pelvis so that it no longer propelled the back legs forward (which is what it does for great apes). Instead, the pelvis had to be re-imagined as a load-bearing device capable of keeping the head above the grass (which is what it does for you). Walking on two legs had several consequences. For one thing, it freed up our hands. For another, it was energy-efficient. It used fewer calories than walking on four legs. Our ancestral bodies used the energy surplus not to pump up our muscles but to pump up our minds—to the point that our modern-day brain, 2 percent of our body weight, sucks up 20 percent of the energy we consume.

These changes in the structure of the brain led to the masterpiece of evolution, the region that distinguishes humans from all other creatures. It is a specialized area of the frontal lobe, just behind the forehead, called the prefrontal cortex.

We got our first hints about its function from a man named Phineas Gage, who suffered the most famous occupational injury in the history of brain science. The injury didn't kill him, but his family probably wished it had. Gage was a popular foreman of a railroad construction crew. He was funny, clever, hardworking, and responsible, the kind of man any dad would be proud to call "son-in-law." On September 13, 1848, he set an explosives charge in the hole of a rock using a tamping iron, a 3-foot rod about an inch in diameter. The charge blew the rod into Gage's head, entering just under the eye and destroying most of his prefrontal cortex. Miraculously, Gage survived, but he became tactless, impulsive, and profane. He left his family and wandered aimlessly from job to job. His friends said he was no longer Gage.

This was the first real evidence that the prefrontal cortex governs several uniquely human cognitive talents, called "executive functions": solving problems, maintaining attention, and inhibiting emotional impulses. In short, this region controls many of the behaviors that separate us from other animals. And from teenagers.

meet your brain

The prefrontal cortex is only the newest addition to the brain. Three brains are tucked inside your head, and parts of their structure took millions of years to design. (This "triune theory of the brain" is one of several models scientists use to describe the brain's overarching structural organization.) Your most ancient neural structure is the brain stem, or "lizard brain." This rather insulting label reflects the fact that the brain stem functions the same in you as in a gila monster. The brain stem controls most of your body's housekeeping chores. Its neurons regulate breathing, heart rate, sleeping, and waking. Lively as Las Vegas, they are always active, keeping your brain purring along whether you're napping or wide awake.

Sitting atop your brain stem is what looks like a sculpture of a scorpion carrying a slightly puckered egg on its back. The Paleomammalian brain appears in you the same way it does in many mammals, such as house cats, which is how it got its name. It has more to do with your animal survival than with your human potential. Most of its functions involve what some researchers call the "four F's": fighting, feeding, fleeing, and ... reproductive behavior.

Several parts of this "second brain" play a large role in the Brain Rules. The claw of the scorpion, called the amygdale, allows you to feel rage. Or fear. Or pleasure. Or memories of past experiences of rage, fear, or pleasure. The amygdala is responsible for both the creation of emotions and the memories they generate. The leg attaching the claw to the body of the scorpion is called the

you have three brains

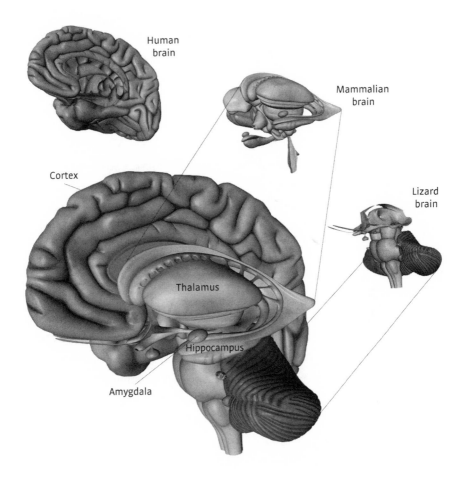

Human
brain

Mammalian
brain

Cortex

Lizard
brain

Thalamus

Hippocampus

Amygdala

More illustrations at www.brainrules.net

hippocampus. The hippocampus converts your short-term memories into longer-term forms. The scorpion's tail curls over the egg-shaped structure like the letter "C," as if protecting it. This egg is the thalamus, one of the most active, well-connected parts of the brain—a control tower for the senses. Sitting squarely in the center of your brain, it processes signals sent from nearly every corner of your sensory universe, then routes them to specific areas throughout your brain.

How this happens is mysterious. Large neural highways run overhead these two brains, combining with other roads, branching suddenly into thousands of exits, bounding off into the darkness. Neurons spark to life, then suddenly blink off, then fire again. Complex circuits of electrical information crackle in coordinated, repeated patterns, then run off into the darkness, communicating their information to unknown destinations.

Arching above like a cathedral is your "human brain," the cortex. Latin for "bark," the cortex is the surface of your brain. It is in deep electrical communication with the interior. This "skin" ranges in thickness from that of blotting paper to that of heavy-duty cardboard. It appears to have been crammed into a space too small for its surface area. Indeed, if your cortex were unfolded, it would be about the size of a baby blanket.

It looks monotonous, slightly like the shell of a walnut, which fooled anatomists for hundreds of years. Until World War I came along, they had no idea each region of the cortex was highly specialized, with sections for speech, for vision, for memory. World War I was the first major conflict where large numbers of combatants encountered shrapnel, and where medical know-how allowed them to survive their injuries. Some of these injuries penetrated only to the periphery of the brain, destroying tiny regions of cortex while leaving everything else intact. Enough soldiers were hurt that scientists could study in detail the injuries and the truly strange behaviors that resulted. Horribly confirming their findings during World War II, scientists eventually were able to make a complete structure-function map of the brain—and see how it had changed over the eons.

They found that as our brains evolved, our heads did, too: They were getting bigger all the time. Tilted hips and big heads are not easy anatomical neighbors. The pelvis—and birth canal—can be only so wide, which is bonkers if you are giving birth to children with larger and larger heads. A lot of mothers and babies died on the way to reaching an anatomical compromise. Human pregnancies are still remarkably risky without modern medical intervention. The solution? Give birth while the baby's head is small enough to fit through the birth canal. The problem? You create childhood. The brain could conveniently finish its developmental programs outside the womb, but the trade-off was a creature vulnerable to predation for years and not reproductively fit for more than a decade. That's an eternity if you make your living in the great outdoors, and outdoors was our home address for eons. But it was worth it. During this time of extreme vulnerability, you had a creature fully capable of learning just about anything and, at least for the first few years, not good for doing much else. This created the concept not only of learner but, for adults, of teacher. It was in our best interests to teach well: Our genetic survival depended upon our ability to protect the little ones.

Of course, it was no use having babies who took years to grow if the adults were eaten before they could finish their thoughtful parenting. Weaklings like us needed a tactic that could allow us to outcompete the big boys in their home turf, leaving our new home safer for sex and babies. We decided on a strange one. We decided to try to get along with each other.

you scratch my back...

Suppose you are not the biggest person on the block, but you have thousands of years to become one. What do you do? If you are an animal, the most straightforward approach is becoming physically bigger, like the alpha male in a dog pack, with selection favoring muscle and bone. But there is another way to double your biomass. It's not by creating a body but by creating an ally. If you can establish cooperative agreements with some of your neighbors, you can double your power even if you do not personally double your strength. You can dominate the world. Trying to fight off a woolly mammoth? Alone, and the fight might look like Bambi vs. Godzilla. Two or three of you, however, coordinating your behaviors and establishing the concept of "teamwork," and you present a formidable challenge. You can figure out how to compel the mammoth to tumble over a cliff, for one. There is ample evidence that this is exactly what we did.

This changes the rules of the game. We learned to cooperate, which means creating a shared goal that takes into account your allies' interests as well as your own. Of course, in order to understand your allies' interests, you must be able to understand others' motivations, including their reward and punishment systems. You need to know where their "itch" is.

Understanding how parenting and group behavior allowed us to dominate our world may be as simple as understanding a few ideas behind the following sentence: The husband died, and then the wife died. There is nothing particularly interesting about that sentence, but watch what happens when I add two small words at the end: The husband died, and then the wife died of grief. All of a sudden we have a view, however brief, into the psychological interior of the wife. We have an impression of her mental state, perhaps even knowledge about her relationship with her husband.

These inferences are the signature characteristic of something called Theory of Mind. We activate it all the time. We try to see our entire world in terms of motivations, ascribing motivations to our pets and even to inanimate objects. (I once knew a guy who treated his 25-foot sailboat like a second wife. Even bought her gifts!) The skill is useful for selecting a mate, for navigating the day-to-day issues surrounding living together, for parenting. Theory of Mind is something humans have like no other creature. It is as close to mind reading as we are likely to get.

This ability to peer inside somebody's mental life and make predictions takes a tremendous amount of intelligence and, not surprisingly, brain activity. Knowing where to find fruit in the jungle is cognitive child's play compared with predicting and manipulating other people within a group setting. Many researchers believe a direct line exists between the acquisition of this skill and our intellectual dominance of the planet.

When we try to predict another person's mental state, we have physically very little to go on. Signs do not appear above a person's head, flashing in bold letters his or her motivations. We are forced to detect characteristics that are not physically obvious at all. This talent is so automatic, we hardly know when we do it. We began doing it in every domain. Remember the line that we can transform into a "1" and an "i"? Now you have dual representation: the line and the thing the line represents. That means you have Judy DeLoache, and that means you have us. Our intellectual prowess, from language to mathematics to art, may have come from the powerful need to predict our neighbor's psychological interiors.

feeling it

It follows from these ideas that our ability to learn has deep roots in relationships. If so, our learning performance may be deeply affected by the emotional environment in which the learning takes place. There is surprising empirical data to support this. The quality of education may in part depend on the relationship between student and teacher. Business success may in part depend on the relationship between employee and boss.

I remember a story by a flight instructor I knew well. He told me about the best student he ever had, and a powerful lesson he learned about what it meant to teach her. The student excelled in ground school. She aced the simulations, aced her courses. In the skies, she showed surprisingly natural skill, quickly improvising even in rapidly changing weather conditions. One day in the air, the instructor saw her doing something naïve. He was having a bad day and he yelled at her. He pushed her hands away from the airplane's equivalent of a steering wheel. He pointed angrily at an instrument. Dumbfounded, the student tried to correct herself, but in the stress of the moment, she made more errors, said she couldn't think, and then buried her head in her hands and started to cry. The teacher took control of the aircraft and landed it. For a long time, the student would not get back into the same cockpit. The incident hurt not only the teacher's professional relationship with the student but the student's ability to learn. It also crushed the instructor. If he had been able to predict how the student would react to his threatening behavior, he never would have acted that way.

If someone does not feel safe with a teacher or boss, he or she may not be able to perform as well. If a student feels misunderstood because the teacher cannot connect with the way the student learns, the student may become isolated. This lies at the heart of the flight student's failure. As we'll see in the Stress chapter, certain types of learning wither in the face of traumatic stress. As we'll see in the Attention chapter, if a teacher can't hold a student's interest, knowledge will not be richly encoded in the brain's database. As we see in this chapter, relationships matter when attempting to teach human beings. Here we are talking about the highly intellectual venture of flying an aircraft. But its success is fully dependent upon feelings.

It's remarkable that all of this came from an unremarkable change in the weather. But a clear understanding of this affords us our first real insight into how humans acquire knowledge: We learned to improvise off a database, with a growing ability to think symbolically about our world. We needed both abilities to survive on the savannah. We still do, even if we have exchanged it for classrooms and cubicles.

Summary

Rule #2
The human brain evolved, too.

* We don't have one brain in our heads; we have three. We started with a "lizard brain" to keep us breathing, then added a brain like a cat's, and then topped those with the thin layer of Jell-O known as the cortex—the third, and powerful, "human" brain.

* We took over the Earth by adapting to change itself, after we were forced from the trees to the savannah when climate swings disrupted our food supply.

* Going from four legs to two to walk on the savannah freed up energy to develop a complex brain.

* Symbolic reasoning is a uniquely human talent. It may have arisen from our need to understand one another's intentions and motivations, allowing us to coordinate within a group.

Get more at www.brainrules.net

wiring

Rule #3
Every brain is wired differently.

MICHAEL JORDAN'S ATHLETIC FAILURES are puzzling, don't you think?

In 1994, one of the best basketball players in the world—ESPN's greatest athlete of the 20th century— decided to quit the game and take up baseball instead. Jordan failed miserably, hitting .202 in his only full season, the lowest of any regular player in the league that year. He simultaneously committed 11 errors in the outfield, also the league's worst. Jordan's performance was so poor, he couldn't even qualify for a triple-A farm team. Though it seems preposterous that anyone with his physical ability would fail at any athletic activity he put his mind to, the fact that Jordan did not even make the minor leagues is palpable proof that you can.

His failure was that much more embarrassing because another athletic legend, Ken Griffey Jr., was burning up the baseball diamond that same year. Griffey was excelling at all of the skills Jordan seemed to lack—and doing so in the *majors*, thank you. Griffey, then playing for the red-hot Seattle Mariners, maintained this excellence for most of the decade, batting .300 for seven years in the 1990s and at the same time slugging out 422 home runs. He is, at this printing, sixth on the all-time home-runs list.

Like Jordan, Griffey Jr. played in the outfield but, unlike Jordan, he was known for catches so spectacular he seemed to float in the air. Float in the *air*? Wasn't that the space Jordan was accustomed to inhabiting? But the sacred atmosphere of the baseball park refused to budge for Jordan, and he eventually went back to what his brains and muscles did better than anyone else's, creating a legendary sequel to a previously stunning basketball career.

What was going on in the bodies of these two athletes? What is it about their brains' ability to communicate with their muscles and skeletons that made their talents so specialized? It has to do with how their brains were wired. To understand what that means, we will watch what happens in the brain as it is learning, discuss the enormous role of experience in brain development—including how identical twins having an identical experience will not emerge with identical brains—and discover that we each have a Jennifer Aniston neuron. I am not kidding.

fried eggs and blueberries

You have heard since grade school that living things are made of cells, and for the most part, that's true. There isn't much that complex biological creatures can do that doesn't involve cells. You may have little gratitude for this generous contribution to your existence, but your cells make up for the indifference by ensuring that you can't control them. For the most part, they purr and hum behind the scenes, content to supervise virtually everything you will ever experience, much of which lies outside your awareness. Some cells are so unassuming, they find their normal function only after they can't function. The surface of your skin, for example—all 9 pounds of it—literally is deceased. This allows the rest of your cells to support your daily life free of wind, rain, and spilled nacho cheese at a basketball game. It is accurate to say that nearly every inch of your outer physical presentation to the world is dead.

The biological structures of the cells that are alive are fairly easy to understand. Most look just like fried eggs. The white of the egg we call the cytoplasm; the center yolk is the nucleus. The nucleus contains that master blueprint molecule and newly christened patron saint of wrongfully convicted criminals: DNA. DNA possesses genes, small snippets of biological instructions, that guide everything from how tall you become to how you respond to stress. A lot of genetic material fits inside that yolk-like nucleus. Nearly 6 feet of the stuff are crammed into a space that is measured in microns. A micron is $1/25,000^{th}$ of an inch, which means putting DNA into your nucleus is like taking 30 miles of fishing line and stuffing it into a blueberry. The nucleus is a crowded place.

One of the most unexpected findings of recent years is that this DNA, or deoxyribonucleic acid, is not randomly jammed into the nucleus, as one might stuff cotton into a teddy bear. Rather, DNA is folded into the nucleus in a complex and tightly regulated manner. The reason for this molecular origami: cellular career options. Fold the DNA one way and the cell will become a contributing member of your liver. Fold it another way and the cell will become part of your busy bloodstream. Fold it a third way and you get a nerve cell—and the ability to read this sentence.

So what does one of those nerve cells look like? Take that fried egg and smash it with your foot, splattering it across the floor. The resulting mess may look like a many-pointed star. Now take one tip of that star and stretch it out. Way out. Using your thumb, now squish the farthest region of the point you just stretched. This creates a smaller version of that multipronged shape. Two smashed stars separated by a long, thin line. There's your typical nerve. Nerve cells come in many sizes and shapes, but most have this basic framework. The foot-stomped fried-egg splatter is called the nerve's cell body. The many points on the resulting star are called dendrites. The region you stretched out is called an axon, and the smaller, thumb-induced starburst at the farther end of the axon is called the axon terminal.

These cells help to mediate something as sophisticated as human thought. To understand how, we must journey into the Lilliputian world of the neuron, and to do that, I would like to borrow from a movie I saw as a child. It was called *Fantastic Voyage*, written by Harry Kleiner and popularized afterward in a book by legendary science-fiction writer Isaac Asimov. Using a premise best described as *Honey, I Shrunk the Submarine*, the film follows a group of researchers exploring the internal workings of the human body—in a submersible reduced to microscopic size. We are going to enter such a submarine, which will allow us to roam around the insides of a typical nerve cell and the watery world in which it is anchored. Our initial port of call is to a neuron that resides in the hippocampus.

When we arrive at this hippocampal neuron, it looks as if we've landed in an ancient, underwater forest. Somehow it has become electrified, which means we are going to have to be careful. Everywhere there are submerged jumbles of branches, limbs, and large, trunk-like objects. And everywhere sparks of electric current run up and down those trunks. Occasionally, large clouds of tiny chemicals erupt from one end of the tree trunks, after the electricity has convulsed through them.

These are not trees. These are neurons, with some odd structural distinctions. Hovering close to one of them, for example, we realize that the "bark" feels surprisingly like grease. That's because it *is* grease. In the balmy interior of the human body, the exterior of the neuron, the phospholipid bilayer, is the consistency of Mazola oil. It's the interior structures that give a neuron its shape, much as the human skeleton gives the body its shape. When we plunge into the interior of the cell, one of the first things we will see is this skeleton.

So let's plunge.

It's instantly, insufferably overcrowded, even hostile, in here. Everywhere we have to navigate through a dangerous scaffolding of spiky, coral-like protein formations: the neural skeleton. Though these dense formations give the neuron its three-dimensional shape, many of the skeletal parts are in constant motion—which means we have to do a lot of dodging. Millions of molecules still slam against our ship, however, and every few seconds we are jolted by electrical discharges. We don't want to stay long.

swimming laps

We escape from one end of the neuron. Instead of perilously winding through sharp thickets of proteins, we now find ourselves free-floating in a calm, seemingly bottomless watery canyon. In the distance, we can see another neuron looming ahead.

We are in the space between the two neurons, called a synaptic cleft, and the first thing we notice is that we are not alone. We appear to be swimming with large schools of tiny molecules. They are streaming out of the neuron we just visited and thrashing helter-skelter toward the one we are facing. In a few seconds, they reverse themselves, swimming back to the neuron we just left. It instantly gobbles them up. These schools of molecules are called neurotransmitters, and they come in a variety of molecular species. They function like tiny couriers, and neurons use these molecules to communicate information across the canyon (or, more properly, the synaptic cleft). The cell that lets them escape is called the pre-synaptic neuron, and the cell that receives them is called the post-synaptic neuron.

Neurons release these chemicals into the synapse usually in response to being electrically stimulated. The neuron that receives them can react negatively or positively when it encounters these chemicals. Working something like a cellular temper tantrum, the neuron can turn itself off to the rest of the neuroelectric world (a process termed inhibition). Or, the neuron can become electrically stimulated. That allows a signal to be transferred from the pre-synaptic neuron to the post: "I got stimulated and I am passing on

55

the good news to you." Then the neurotransmitters return to the cell of origin, a process appropriately termed re-uptake. When that cell gobbles them up, the system is reset and ready for another signal.

As we look 360 degrees around our synaptic environment, we notice that the neural forest, large and seemingly distant, is surprisingly complicated. Take the two neurons between which we are floating. We are between just two connection points. If you can imagine two trees being uprooted by giant hands, turned 90 degrees so the roots face each other, and then jammed together, you can visualize the real world of two neurons interacting with each other in the brain. And that's just the simplest case. Usually, thousands of neurons are jammed up against one another, all occupying a single small parcel of neural real estate. The branches form connections to one another in a nearly incomprehensible mass of branching confusion. Ten thousand points of connection is typical, and each connection is separated by a synapse, those watery canyons in which we are now floating.

Gazing at this underwater hippocampal forest, we notice several disturbing developments. Like snakes swaying to the rhythm of some chemical flute, some of these branches appear to be moving. Occasionally, the end of one neuron swells up, greatly increasing in diameter. The terminal ends of other neurons split down the middle like a forked tongue, creating two connections where there was only one. Electricity crackles through these moving neurons at a blinding 250 miles per hour, some quite near us, with clouds of neurotransmitters filling the spaces between the trunks as the electric current passes by.

What we should do now is take off our shoes and bow low in the submarine, for we are on Holy Neural Ground. What we are observing is the process of the human brain *learning*.

extreme makeover

Eric Kandel is the scientist mostly responsible for figuring out the cellular basis of this process. For it, he shared the Nobel Prize in 2000, and his most important discoveries would have made inventor Alfred Nobel proud. Kandel showed that when people learn something, the wiring in their brains changes. He demonstrated that acquiring even simple pieces of information involves the physical alteration of the structure of the neurons participating in the process. Taken broadly, these physical changes result in the functional organization and reorganization of the brain. This is astonishing. The brain is constantly learning things, so the brain is constantly rewiring itself.

Kandel first discovered this fact not by looking at humans but by looking at sea slugs. He soon found, somewhat insultingly, that human nerves learn things in the same way slug nerves learn things. And so do lots of animals in between slugs and humans. The Nobel Prize was awarded in part because his careful work described the thought processes of virtually every creature with the means to think.

We saw these physical changes while our submarine was puttering around the synaptic space between two neurons. As neurons learn, they swell, sway, and split. They break connections in one spot, glide over to a nearby region, and form connections with their new neighbors. Many stay put, simply strengthening their electrical connections with each other, increasing the efficiency of information transfer. You can get a headache just thinking about the fact that deep inside your brain, at this very moment, bits of neurons are moving around like reptiles, slithering to new spots, getting fat at one end or creating split ends. All so that you can remember a few things about Eric Kandel and his sea slugs.

But before Kandel, in the 18th century, the Italian scientist Vincenzo Malacarne did a surprisingly modern series of biological experiments. He trained a group of birds to do complex tricks, killed them all, and dissected their brains. He found that his trained birds had more extensive folding patterns in specific regions of their brains than his untrained birds. Fifty years later, Charles Darwin noted similar differences between the brains of wild animals and their domestic counterparts. The brains in wild animals were 15 to 30 percent larger than those of their tame, domestic counterparts. It appeared that the cold, hard world forced the wild animals into a constant learning mode. Those experiences wired their heads much differently.

It is the same with humans. This can be observed in places ranging from New Orleans's Zydeco beer halls to the staid palaces of the New York Philharmonic. Both are the natural habitat of violin players, and violin players have really strange brains when compared with non-violin players. The neural regions that control their left hands, where complex, fine motor movement is required on the strings, look as if they've been gorging on a high-fat diet. These regions are enlarged, swollen and crisscrossed with complex associations. By contrast, the areas controlling the right hand, which draws the bow, looks positively anorexic, with much less complexity.

The brain acts like a muscle: The more activity you do, the larger and more complex it can become. Whether that leads to more intelligence is another issue, but one fact is indisputable: What you do in life physically changes what your brain looks like. You can wire and rewire yourself with the simple choice of which musical instrument—or professional sport—you play.

some assembly required

How does this fantastic biology work? Infants provide a front-row seat to one of the most remarkable construction projects on Earth. Every newly born brain should come with a sticker saying "some assembly required." The human brain, only partially constructed at birth, won't be fully assembled for years to come. The biggest construction programs aren't finished until you are in your early 20s, with fine-tuning well into your mid-40s.

When babies are born, their brains have about the same number of connections as adults have. That doesn't last long. By the time children are 3 years old, the connections in specific regions of their brains have doubled or even tripled. (This has given rise to the popular belief that infant brain development is *the* critical key to intellectual success in life. That's not true, but that's another story.) This doubling and tripling doesn't last long, either. The brain soon takes thousands of tiny pruning shears and trims back a lot of this hard work. By the time children are 8 or so, they're back to their adult numbers. And if kids never went through puberty, that would be the end of the story. In fact, it is only the middle of the story.

At puberty, the whole thing starts over again. Quite different regions in the brain begin developing. Once again, you see frenetic neural outgrowth and furious pruning back. It isn't until parents begin thinking about college financial aid for their high schoolers that their brains begin to settle down to their adult forms (sort of). It's like a double-humped camel. From a connectivity point of view, there is a great deal of activity in the terrible twos and then, during the terrible teens, a great deal more.

Though that might seem like cellular soldiers obeying growth commands in lockstep formation, nothing approaching military precision is observed in the messy world of brain development. And it is at this imprecise point that brain development meets Brain Rule. Even a cursory inspection of the data reveals remarkable variation in growth patterns from one person to the next. Whether examining toddlers or teenagers, different regions in different children develop at different rates. There is a remarkable degree of diversity in the specific areas that grow and prune, and with what enthusiasm they do so.

I'm reminded of this whenever I see the class pictures that captured my wife's journey through the American elementary-school system. My wife went to school with virtually the same people for her entire K–12 experience (and actually remained friends with most of them). Though the teachers' dated hairstyles are the subject of much laughter for us, I often focus on what the kids looked like back then. I always shake my head in disbelief.

In the first picture, the kids are all in grade one. They're about the same age, but they don't look it. Some kids are short. Some are tall. Some look like mature little athletes. Some look as if they just got out of diapers. The girls almost always appear older than the boys. It's even worse in the junior-high pictures of this same class. Some of the boys look as if they haven't developed much since third grade. Others are clearly beginning to sprout whiskers. Some of the girls, flat chested and uncurved, look a lot like boys. Some seem developed enough to make babies.

Why do I bring this up? If we had X-ray eyes capable of penetrating their little skulls, we would find that the brains of these kids are *just as unevenly developed* as their bodies.

the jennifer aniston neuron

We are born into this world carrying a number of preset circuits. These control basic housekeeping functions like breathing, heartbeat, your ability to know where your foot is even if you can't see it, and so on. Researchers call this "experience independent" wiring. The brain also leaves parts of its neural construction project unfinished at birth, waiting for external experience to direct it. This "experience expectant" wiring is related to areas such as visual acuity and perhaps language acquisition. And, finally, we have "experience dependent" wiring. It may best be explained with a story about Jennifer Aniston. You might want to skip the next paragraph if you are squeamish.

Ready? A man is lying in surgery with his brain partially exposed to the air. He is conscious. The reason he is not crying out in agony is that the brain has no pain neurons. He can't feel the needle-sharp electrodes piercing his nerve cells. The man is about to have some of his neural tissue removed—resected, in surgical parlance— because of intractable, life-threatening epilepsy. Suddenly, one of the surgeons whips out a photo of Jennifer Aniston and shows it to the patient. A neuron in the man's head fires excitedly. The surgeon lets out a war whoop.

Sound like a grade B movie? This experiment really happened. The neuron in question responded to seven photographs of actress Jennifer Aniston, while it practically ignored the 80 other images of everything else, including famous and non-famous people. Lead scientist Quian Quiroga said, "The first time we saw a neuron firing to seven different pictures of Jennifer Aniston—and nothing else— we literally jumped out of our chairs." There is a neuron lurking in your head that is stimulated only when Jennifer Aniston is in the room.

A Jennifer Aniston *neuron*? How could this be? Surely there is nothing in our evolutionary history suggesting that Jennifer Aniston is a permanent denizen of our brain wiring. (Aniston wasn't even born until 1969, and there are regions in our brain whose designs are millions of years old). To make matters worse, the researchers also found a Halle Berry-specific neuron, a cell in the man's brain that wouldn't respond to pictures of Aniston or anything else. Just Berry. He also had a neuron specific to Bill Clinton. It no doubt was helpful to have a sense of humor while doing this kind of brain research.

Welcome to the world of experience-dependent brain wiring, where a great deal of the brain is hard-wired *not* to be hard-wired. Like a beautiful, rigorously trained ballerina, we are hard-wired to be flexible.

We can immediately divide the world's brains into those who know of Jennifer Aniston or Halle Berry and those who don't. The brains of those who do are not wired the same way as those who don't. This seemingly ridiculous observation underlies a much larger concept. Our brains are so sensitive to external inputs that their physical wiring depends upon the culture in which they find themselves.

Even identical twins do not have identical brain wiring. Consider this thought experiment: Suppose two adult male twins rent the

Halle Berry movie *Catwoman*, and we in our nifty little submarine are viewing their brains while they watch. Even though they are in the same room, sitting on the same couch, the twins see the movie from slightly different angles. We find that their brains are encoding visual memories of the video differently, in part because it is impossible to observe the video from the same spot. Seconds into the movie, they are already wiring themselves differently.

One of the twins earlier in the day read a magazine story about panned action movies, a picture of Berry figuring prominently on the cover. While watching the video, this twin's brain is simultaneously accessing memories of the magazine. We observe that his brain is busy comparing and contrasting comments from the text with the movie and is assessing whether he agrees with them. The other twin has not seen this magazine, so his brain isn't doing this. Even though the difference may seem subtle, the two brains are creating different memories of the same movie.

That's the power of the Brain Rule. Learning results in physical changes in the brain, and these changes are unique to each individual. Not even identical twins having identical experiences possess brains that wire themselves exactly the same way. And you can trace the whole thing to experience.

on the street where you live

Perhaps a question is now popping up in your brain: If every brain is wired differently from every other brain, can we know *anything* about the organ?

Well, yes. The brain has billions of cells whose collective electrical efforts make a loving, wonderful you or, perhaps with less complexity, Kandel's sea slug. All of these nerves work in a similar fashion. Every human comes equipped with a hippocampus, a pituitary gland, and the most sophisticated thinking store of electrochemistry on the planet: a cortex. These tissues function the same way in every brain.

How then can we explain the individuality? Consider a highway.

The United States has one of the most extensive and complex ground transportation systems in the world. There are lots of variations on the idea of "road," from interstate freeways, turnpikes, and state highways to residential streets, one-lane alleys, and dirt roads. Pathways in the human brain are similarly diverse. We have the neural equivalents of large interstate freeways, turnpikes, and state highways. These big trunks are the same from one person to the next, functioning in yours about the same way they function in mine. So a great deal of the structure and function of the brain is predictable, a property that allows the word "science" to be attached to the end of the word "neuro" and keeps people like me employed. Such similarity may be the ultimate fruit of the double-humped developmental program we talked of previously. That's the experience-independent wiring.

It's when you get to the smaller routes—the brain's equivalent of residential streets, one-laners and dirt roads—that individual patterns begin to show up. Every brain has a lot of these smaller paths, and in no two people are they identical. The individuality is seen at the level of the very small, but because we have so much of it, the very small amounts to a big deal.

It is one thing to demonstrate that every brain is wired differently from every other brain. It is another to say that this affects intelligence. Two scientists, a behavioral theorist and a neurosurgeon, offer differing perspectives on the subject. The theorist believes in seven to nine categories of multiple intelligence. The neurosurgeon also believes in multiple categories. He thinks there may be billions.

Meet Howard Gardner, psychologist, author, educator, and father of the so-called Multiple Intelligences movement. Gardner had the audacity to suggest that the competency of the human mind is too multifaceted to be boiled down to simplistic numerical measures. He threw out the idea of IQ tests, and then he attempted to reframe the question of human intellectual skill. Like a cognitive Jane Goodall in an urban jungle, Gardner and his colleagues observed real people in the act of learning—at school, at work, at play, at *life*. He began to notice categories of intellectual talent that people used every day that were not always identified as being "intelligent" and certainly were not measurable by IQ tests. After thinking about things for a long time, he published his findings in a book called *Frames of Mind: The Theory of Multiple Intelligences*. It set off a firestorm of debate that burns unabated to this day.

Gardner believes he has observed at least seven categories of intelligence: verbal/linguistic, musical/rhythmic, logical/mathematical, spatial, bodily/kinesthetic, interpersonal, and intrapersonal. He calls these "entry points" into the inner workings of the human mind. The categories don't always intersect with one another, and Gardner has said, "If I know you're very good in music, I can predict with just about zero accuracy whether you're going to be good or bad in other things."

Some researchers think Gardner is resting on his opinion, not on his data. But none of his critics attack the underlying thesis that the human intellect is multifaceted. To date, Gardner's efforts represent the first serious attempt to provide an alternative to numerical descriptions of human cognition.

mapping the brain

But categories of intelligence may number more than 7 billion—roughly the population of the world. You can get a sense of this by watching skilled neurosurgeon George Ojemann examine the exposed brain of a 4-year-old girl. Ojemann has a shock of white hair, piercing eyes, and the quiet authority of someone who for decades has watched people live and die in the operating room. He is one of the great neurosurgeons of our time, and he is an expert at a technique called electrical stimulation mapping.

He is hovering over a girl with severe epilepsy. She is fully conscious, her brain exposed to the air. He is there to remove some of her misbehaving brain cells. Before Ojemann takes out anything, however, he has to make a map. He wields a slender white wand attached to a wire, a cortical stimulator, which sends out small, unobtrusive electrical shocks to anything it touches. If it brushed against your hand, you would feel only a slight tingly sensation.

Ojemann gently touches one end of the wand to an area of the little girl's brain and then asks her, "Did you feel anything?" She says dreamily, "Somebody just touched my hand." He puts a tiny piece of paper on the area. He touches another spot. She exclaims, "Somebody just touched my cheek!" Another tiny piece of paper. This call and response goes on for hours. Like a neural cartographer, Ojemann is mapping the various functions of his little patient's brain, with special attention paid to the areas close to her epileptic tissue.

These are tests of the little girl's motor skills. For reasons not well understood, however, epileptic tissues are often disturbingly adjacent to critical language areas. So Ojemann also pays close attention to the regions involved in language processing, where words and sentences and grammatical concepts are stored. This child happens to be bilingual, so language areas essential for both Spanish and English will need to be mapped. A paper dot marked "S" is applied to the regions where Spanish exists, and a small "E" where English is stored. Ojemann does this painstaking work with every single patient who undergoes this type of surgery. Why? The answer is a stunner. He has to map each individual's critical function areas because *he doesn't know where they are.*

Ojemann can't predict the function of very precise areas in advance of the surgery because no two brains are wired identically. Not in terms of structure. Not in terms of function. For example, from nouns to verbs to aspects of grammar, we each store language in different areas, recruiting different regions for different components. Bilingual people don't even store their Spanish and their English in similar places.

This individuality has fascinated Ojemann for years. He once combined the brain maps for 117 patients he had operated on over the years. Only in one region did he find a spot where most people had a critical language area, or CLA, and "most" means 79 percent of the patients.

Data from electrical stimulation mapping give probably the most dramatic illustration of the brain's individuality. But Ojemann also wanted to know how stable these differences were during life, and if any of those differences predicted intellectual competence. He found interesting answers to both questions. First, the maps are established very early in life, and they remain stable throughout. Even if a decade or two had passed between surgeries, the regions recruited for a specific CLA remained recruited for that same CLA. Ojemann also found that certain CLA patterns could predict language competency, at least as measured by a pre-operative verbal IQ test. If you want to be good at a language (or at least perform well on the test), don't let the superior temporal gyrus host your CLA. Your verbal performance will statistically be quite poor. Also, make sure your overall CLA pattern has a small and rather tightly focused footprint. If the pattern is instead widely distributed, you will have a remarkably low score. These findings are robust and age-independent. They have been demonstrated in people as young as kindergartners and as old as Alan Greenspan.

Not only are people's brains individually wired, but those neurological differences can, at least in the case of language, predict performance.

ideas

Given these data, does it make any sense to have school systems that expect every brain to learn like every other? Does it make sense to treat everybody the same in business, especially in a global economy replete with various cultural experiences? The data offer powerful implications for how we should teach kids—and, when they grow up and get a job, how we should treat them as employees. I have a couple of concerns about our school system:

1) The current system is founded on a series of expectations that certain learning goals should be achieved by a certain age. Yet there is no reason to suspect that the brain pays attention to those expectations. Students of the *same age* show a great deal of intellectual variability.

2) These differences can profoundly influence classroom performance. This has been tested. For example, about 10 percent of students do *not* have brains sufficiently wired to read at the age at which we expect them to read. Lockstep models based simply on age are guaranteed to create a counterproductive mismatch to brain biology.

What can we do about this?

Smaller class size

All else being equal, it has been known for many years that smaller, more intimate schools create better learning environments than megaplex houses of learning. The Brain Rule may help explain why smaller is better.

Given that every brain is wired differently, being able to read a student's mind is a powerful tool in the hands of a teacher. As you may recall from the Survival chapter, Theory of Mind is about as close to mind reading as humans are likely to get. It is defined as the ability to understand the interior motivations of someone else and the ability to construct a predictable "theory of how their mind works" based on that knowledge. This gives teachers critical access to their students' interior educational life. It may include knowledge of when students are confused and when they are fully engaged. It also gives sensitive teachers valuable feedback about whether their teaching is being transformed into learning. It may even be the definition of that sensitivity. I have come to believe that people with advanced Theory of Mind skills possess the single most important ingredient for becoming effective communicators of information.

Students comprehend complex knowledge at different times and at different depths. Because a teacher can keep track of only so many minds, there must be a limit on the number of students in a class—the smaller, the better. It is possible that small class sizes predict better performance simply because the teacher can better keep track of where everybody is. This suggests that an advanced skill set in Theory of Mind predicts a good teacher. If so, existing Theory of Mind tests could be used like Myers-Briggs personality tests to reveal good teachers from bad, or to help people considering careers as teachers.

Customized instruction

What of that old admonition to create more individualized instruction within a grade level? It sits on some solid brain science. Researcher Carol McDonald Connor is doing the first work I've seen capable of handling these differences head-on. She and a colleague combined a standard reading program with a bright and shiny new computer program called A2i. The software uses artificial intelligence to determine where the user's reading competencies lie and then adaptively tailor exercises for the student in order to fill in any gaps.

When used in conjunction with a standard reading class, the software is wildly successful. The more students work with the program, the better their scores become. Interestingly, the effect is greatest when the software is used in conjunction with a normal reading program. Teacher alone or software alone is not as effective. As the instructor teaches the class in a normal fashion, students will, given the uneven intellectual landscape, experience learning gaps. Left untreated, these gaps cause students to fall further and further behind, a normal and insidious effect of not being able to transform instruction into apprehension. The software makes sure these gaps don't go untreated.

Is this the future? Attempting to individualize education is hardly a new idea. Using code as a stand-in for human teaching is not revolutionary, either. But the combination might be a stunner. I would like to see a three-pronged research effort between brain and education scientists:

1) Evaluate teachers and teachers-to-be for advanced Theory of Mind skills, using one of the four main tests that measure empathy. Determine whether this affects student performance in a statistically valid fashion.

2) Develop adaptive software for a variety of subjects and grade levels. Test them for efficacy. Deploy the ones that work in a manner similar to the experiment Connor published in the journal *Science*.

3) Test both ideas in various combinations. Add to the mix environments where the student-teacher ratio is both typical and optimized, and then compare the results.

The reason to do this is straightforward: You cannot change the fact that the human brain is individually wired. Every student's brain, every employee's brain, every customer's brain is wired differently. That's the Brain Rule. You can either accede to it or ignore it. The current system of education chooses the latter, to our detriment. It needs to be torn down and newly envisioned, in a Manhattan Project-size commitment to individualizing instruction. We might, among other things, dismantle altogether grade structures based on age.

Companies could try Theory of Mind screening for leaders, along with a method of "mass customization" that treats every employee as an individual. I bet many would discover that they have a great basketball player in their organization, and they're asking him or her to play baseball.

Summary

Rule #3
Every brain is wired differently.

* What you do and learn in life physically changes what your brain looks like—it literally rewires it.

* The various regions of the brain develop at different rates in different people.

* No two people's brains store the same information in the same way in the same place.

* We have a great number of ways of being intelligent, many of which don't show up on IQ tests.

Get more at www.brainrules.net

! attention

Rule #4
We don't pay attention to boring things.

O IT WAS ABOUT 3 o'clock in the morning when I suddenly was startled into consciousness by the presence of a small spotlight sweeping across the walls of our living room. In the moonlight, I could see the 6-foot frame of a young man in a trenchcoat, clutching a flashlight and examining the contents of our house. His other hand held something metallic, glinting in the silvery light. As my sleepy brain was immediately and violently aroused, it struck me that my home was about to be robbed by someone younger than me, bigger than me, and in possession of a firearm. Heart pounding, knees shaking, I crept to the phone, quickly called the police, turned on the lights, went to stand guard outside my children's room, and prayed. Miraculously, a police car was already in the vicinity and activated its sirens within a minute of my phone call. This all happened so quickly that my would-be assailant left his get-away car in our driveway, engine still running. He was quickly apprehended.

That experience lasted only 45 seconds, but aspects of it are indelibly impressed in my memory, from the outline of the young man's coat to the shape of his firearm.

Does it matter to learning if we pay attention? The short answer is: You bet it does. My brain fully aroused, I will never forget that experience as long as I live. The more attention the brain pays to a given stimulus, the more elaborately the information will be encoded—and retained. That has implications for your employees, your students, and your kids. A strong link between attention and learning has been shown in classroom research both a hundred years ago and as recently as last week. The story is consistent: Whether you are an eager preschooler or a bored-out-of-your-mind undergrad, better attention always equals better learning. It improves retention of reading material, accuracy, and clarity in writing, math, science—every academic category that has ever been tested.

So I ask this question in every college course I teach: "Given a class of medium interest, not too boring and not too exciting, when do you start glancing at the clock, wondering when the class will be over?" There is always some nervous shuffling, a few smiles, then a lot of silence. Eventually someone blurts out:

"Ten minutes, Dr. Medina."

"Why 10 minutes?" I inquire.

"That's when I start to lose attention. That's when I begin to wonder when this torment will be over." The comments are always said in frustration. A college lecture is still about 50 minutes long.

Peer-reviewed studies confirm my informal inquiry: Before the first quarter-hour is over in a typical presentation, people *usually* have checked out. If keeping someone's interest in a lecture were a business, it would have an 80 percent failure rate. What happens at the 10-minute mark to cause such trouble? Nobody knows. The brain seems to be making choices according to some stubborn timing pattern, undoubtedly influenced by both culture and gene. This fact suggests a teaching and business imperative: Find a way to arouse and then hold somebody's attention for a specific period of time. But how? To answer that question, we will need to explore some complex pieces of neurological real estate. We are about to investigate the remarkable world of human attention—including what's going on in our brains when we turn our attention to something, the importance of emotions, and the myth of multitasking.

can i have your attention, please?

While you are reading this paragraph, millions of sensory neurons in your brain are firing simultaneously, all carrying messages, each attempting to grab your attention. Only a few will succeed in breaking through to your awareness, and the rest will be ignored either in part or in full. Incredibly, it is easy for you to alter this balance, effortlessly granting airplay to one of the many messages you were previously ignoring. (While still reading this sentence, can you feel where your elbows are right now?) The messages that do grab your attention are connected to memory, interest, and awareness.

memory

What we pay attention to is often profoundly influenced by memory. In everyday life, we use previous experience to predict where we should pay attention. Different environments create different expectations. This was profoundly illustrated by the scientist Jared Diamond in his book *Guns, Germs, and Steel.* He describes an adventure traipsing through the New Guinea jungle with native New Guineans. He relates that these natives tend to perform poorly at tasks Westerners have been trained to do since childhood. But they are hardly stupid. They can detect the most subtle changes in the jungle, good for following the trail of a predator or for finding the way back home. They know which insects to leave alone, know where food exists, can erect and tear down shelters with ease. Diamond, who had never spent time in such places, has no ability to pay attention to these things. Were he to be tested on such tasks, he also would perform poorly.

Culture matters, too, even when the physical ecologies are similar. For example, urban Asians pay a great deal of attention to the context of a visual scene and to the relationships between foreground objects and backgrounds. Urban Americans don't. They pay attention to the focal items before the backgrounds, leaving perceptions of context much weaker. Such differences can affect how an audience perceives a given business presentation or class lecture.

interest

Happily, there are some commonalities regardless of culture. For example, we have known for a long time that "interest" or "importance" is inextricably linked to attention. Researchers sometimes call this arousal. Exactly how it relates to attention is still a mystery. Does interest create attention? We know that the brain continuously scans the sensory horizon, with events constantly assessed for their potential interest or importance. The more important events are then given extra attention. Can the reverse occur, with attention creating interest?

Marketing professionals think so. They have known for years that novel stimuli—the unusual, unpredictable, or distinctive—are powerful ways to harness attention in the service of interest. One well-known example is a print ad for Sauza Conmemorativo tequila. It shows a single picture of an old, dirty, bearded man, donning a brimmed hat and smiling broadly, revealing a single tooth. Printed above the mouth is: "This man only has one cavity." A larger sentence below says: "Life is harsh. Your tequila shouldn't be." Flying in the face of most tequila marketing strategies, which consist of scantily clad 20-somethings dancing at a party, the ad is effective at using attention to create interest.

awareness

Of course, we must be aware of something for it to grab our attention. You can imagine how tough it is to research such an ephemeral concept. We don't know the neural location of consciousness, loosely defined as that part of the mind where awareness resides. (The best data suggest that several systems are scattered throughout the brain.) We have a long way to go before we fully understand the biology behind attention.

One famous physician who has examined awareness at the clinical level is Dr. Oliver Sacks, a delightful British neurologist and one terrific storyteller. One of his most intriguing clinical cases was first described in his bestselling book *The Man Who Mistook His Wife for a Hat*. Sacks describes a wonderful older woman in his care, intelligent, articulate, and gifted with a sense of humor. She suffered a massive stroke in the back region of her brain that left her with a most unusual deficit: She lost the ability to pay attention to anything that was to her left. She could pick up objects only in the right half of her visual field. She could put lipstick only on the right half of her face. She ate only from the right half of her plate. This caused her to complain to the hospital nursing staff that her portions were too small! Only when the plate was turned and the food entered her right visual field could she pay any attention to it and have her fill.

Data like these are very useful to both clinicians and scientists. When damage occurs to a specific brain region, we know that any observed behavioral abnormality must in some way be linked to that region's function. Examining a broad swath of patients like Sacks's gave scientists a cumulative view of how the brain pays attention to things. The brain can be divided roughly into two hemispheres of unequal function, and patients can get strokes in either. Marcel Mesulam of Northwestern University found that the hemispheres contain separate "spotlights" for visual attention. The left hemisphere's spotlight is small, capable of paying attention only to items on the right side of the visual field. The right hemisphere, however, has a global spotlight. According to Mesulam, getting a stroke on the left side is much less catastrophic because the right side can pitch in under duress to aid vision.

Of course, sight is only one stimulus to which the brain is capable of paying attention. Just let a bad smell into the room for a moment or make a loud noise and people easily will shift attention. We also pay close attention to our psychological interiors, mulling over internal events and feelings again and again with complete focus, with no obvious external sensory stimulation. What's going on in our heads when we turn our attention to something?

red alert

Thirty years ago, a scientist by the name of Michael Posner derived a theory about attention that remains popular today. Posner started his research career in physics, joining the Boeing Aircraft Company soon out of college. His first major research contribution was to figure out how to make jet-engine noise less annoying to passengers riding in commercial airplanes. You can thank your relatively quiet airborne ride, even if the screaming turbine is only a few feet from your eardrums, in part on Posner's first research efforts. His work on planes eventually led him to wonder how the brain processes information of any kind. This led him to a doctorate in research and to a powerful idea. Sometimes jokingly referred to as the Trinity Model, Posner hypothesized that we pay attention to things because of the existence of three separable but fully integrated systems in the brain.

One pleasant Saturday morning, my wife and I were sitting on our outdoor deck, watching a robin drink from our birdbath, when all of a sudden we heard a giant "swoosh" above our heads. Looking up, we caught the shadow of a red-tailed hawk, dropping like a thunderbolt from its perch in a nearby tree, grabbing the helpless robin by the throat. As the raptor swooped by us, not 3 feet away, blood from the robin splattered on our table. What started as a leisurely repast ended as a violent reminder of the savagery of the real world. We were stunned into silence.

In Posner's model, the brain's first system functions much like the two-part job of a museum security officer: surveillance and alert. He called it the Alerting or Arousal Network. It monitors the sensory environment for any unusual activities. This is the general level of attention our brains are paying to our world, a condition termed Intrinsic Alertness. My wife and I were using this network as we sipped our coffee, watching the robin. If the system detects something unusual, such as the hawk's swoosh, it can sound an alarm heard brain-wide. That's when Intrinsic Alertness transforms into specific attention, called Phasic Alertness.

After the alarm, we orient ourselves to the attending stimulus, activating the second network. We may turn our heads toward the stimulus, perk up our ears, perhaps move toward (or away) from something. It's why both my wife and I immediately lifted our heads away from the robin, attending to the growing shadow of the hawk. The purpose is to gain more information about the stimulus, allowing the brain to decide what to do. Posner termed this the Orienting Network.

The third system, the Executive Network, controls the "oh my gosh what should I do now" behaviors. These may include setting priorities, planning on the fly, controlling impulses, weighing the consequences of our actions, or shifting attention. For my wife and me, it was stunned silence.

So we have the ability to detect a new stimulus, the ability to turn toward it, and the ability to decide what to do based on its nature. Posner's model offered testable predictions about brain function and attention, leading to neurological discoveries that would fill volumes. Hundreds of behavioral characteristics have since been discovered as well. Four have considerable practical potential: emotions, meaning, multitasking, and timing.

1) Emotions get our attention

Emotionally arousing events tend to be better remembered than neutral events.

While this idea may seem intuitively obvious, it's frustrating to demonstrate scientifically because the research community is still debating exactly what an emotion is. One important area of research is the effect of emotion on learning. An emotionally charged event (usually called an ECS, short for emotionally competent stimulus) is the best-processed kind of external stimulus ever measured. Emotionally charged events persist much longer in our memories and are recalled with greater accuracy than neutral memories.

This characteristic has been used to great effect, and sometimes with great controversy, in television advertising. Consider a television advertisement for the Volkswagen Passat. The commercial opens with two men talking in a car. They are having a mildly heated discussion about one of them overusing the word "like" in conversation. As the argument continues, the viewer notices out the passenger window another car barreling toward the men. It smashes into them. There are screams, sounds of shattering glass, quick-cut shots showing the men bouncing in the car, twisted metal. The exit shot shows the men standing, in disbelief, outside their wrecked Volkswagen. In a twist on a well-known expletive, these words flash on the screen: "Safe Happens." The spot ends with a picture of another Passat, this one intact and complete with its five-star side-crash safety rating. It is a memorable, even disturbing, 30-second spot. And it has these characteristics because its centerpiece is an ECS.

How does this work in our brains? It involves the prefrontal cortex, that uniquely human part of the brain that governs "executive functions" such as problem-solving, maintaining attention, and inhibiting emotional impulses. If the prefrontal cortex is the board chairman, the cingulate gyrus is its personal assistant. The assistant provides the chairman with certain filtering functions and assists in teleconferencing with other parts of the brain—especially the amygdala, which helps create and maintain emotions. The amygdala is chock-full of the neurotransmitter dopamine, and it uses dopamine the way an office assistant uses Post-It notes. When the brain detects an emotionally charged event, the amygdala releases dopamine into the system. Because dopamine greatly aids memory and information processing, you could say the Post-It note reads "Remember this!" Getting the brain to put a chemical Post-It note on a given piece of information means that information is going to be more robustly processed. It is what every teacher, parent, and ad executive wants.

Emotionally charged events can be divided into two categories: those that no two people experience identically, and those that everybody experiences identically.

When my mother got angry (which was rare), she went to the kitchen, washing LOUDLY any dishes she discovered in the sink. And if there were pots and pans, she deliberately would crash them together as she put them away. This noise served to announce to the entire household (if not the city block) her displeasure at something. To this day, whenever I hear loudly clanging pots and pans, I experience an emotionally competent stimulus—a fleeting sense of "You're in trouble now!" My wife, whose mother never displayed anger in this fashion, does not associate anything emotional with the noise of pots and pans. It's a uniquely stimulated, John-specific ECS.

Universally experienced stimuli come directly from our evolutionary heritage, so they hold the greatest potential for use in teaching and business. Not surprisingly, they follow strict Darwinian lines of threats and energy resources. Regardless of who you are, the brain pays a great deal of attention to these questions:

"Can I eat it? Will it eat me?"

"Can I mate with it? Will it mate with me?"

"Have I seen it before?"

Any of our ancestors who didn't remember threatening experiences thoroughly or acquire food adequately would not live long enough to pass on his genes. The human brain has many dedicated systems exquisitely tuned to reproductive opportunity and to the perception of threat. (That's why the robbery story grabbed your attention—and why I put it at the beginning of this chapter.) We also are terrific pattern matchers, constantly assessing our environment for similarities, and we tend to remember things if we think we have seen them before.

One of the best TV spots ever made used all three principles in an ever-increasing spiral. Stephen Hayden produced the commercial, introducing the Apple computer in 1984. It won every major advertising award that year and set a standard for Super Bowl ads. The commercial opens onto a bluish auditorium filled with robot-like men all dressed alike. In a reference to the 1956 movie *1984*, the men are staring at a screen where a giant male face is spouting off platitude fragments such as "information purification!" and "unification of thought!" The men in the audience are absorbing these messages like zombies. Then the camera shifts to a young woman in gym clothes, sledgehammer in hand, running full tilt toward the auditorium. She is wearing red shorts, the only primary color in the entire commercial. Sprinting down the center aisle, she throws her sledgehammer at the screen containing Big Brother. The screen explodes in a hail of sparks and blinding light. Plain letters flash on the screen: "On January 24th, Apple Computer will introduce Macintosh. And you'll see why 1984 won't be like *1984*."

All of the elements are at work here. Nothing could be more threatening to a country marinated in free speech than George Orwell's *1984* totalitarian society. There is sex appeal, with the revealing gym shorts, but there is a twist. Mac is a female, so-o-o … IBM must be a male. In the female-empowering 1980s, a whopping statement on the battle of the sexes suddenly takes center stage. Pattern matching abounds as well. Many people have read *1984* or seen the movie. Moreover, people who were *really* into computers at the time made the connection to IBM, a company often called Big Blue for its suit-clad sales force.

2) Meaning before details

What most people remember about that commercial is its

emotional appeal rather than every detail. There is a reason for that. The brain remembers the emotional components of an experience better than any other aspect. We might forget minute details of an interstate fender bender, for example, yet vividly recall the fear of trying to get to the shoulder without further mishap.

Studies show that emotional arousal focuses attention on the "gist" of an experience *at the expense* of peripheral details. Many researchers think that's how memory normally works—by recording the gist of what we encounter, not by retaining a literal record of the experience. With the passage of time, our retrieval of gist always trumps our recall of details. This means our heads tend to be filled with generalized pictures of concepts or events, not with slowly fading minutiae. I am convinced that America's love of retrieval game shows such as *Jeopardy!* exists because we are dazzled by the unusual people who can invert this tendency.

Of course, at work and at school, detailed knowledge often is critical for success. Interestingly, our reliance on gist may actually be fundamental to finding a strategy for remembering details. We know this from a fortuitous series of meetings that occurred in the 1980s between a brain scientist and waiter.

Watching J.C. take an order is like watching Ken Jennings play *Jeopardy!* J.C. never writes anything down, yet he never gets the order wrong. As the menu offers more than 500 possible combinations of food (entrees, side dishes, salad dressing, etc.) *per customer*, this is an extraordinary achievement. J.C. has been recorded taking the orders of 20 people consecutively with a zero percent error rate. J.C. worked in a restaurant frequented by University of Colorado brain scientist K. Anders Ericsson. Noticing how unusual J.C.'s skills were, he asked J.C. if he would submit to being studied. The secret of J.C.'s success lay in the deployment of a powerful organization strategy. He always divided the customer's order into discrete categories, such as entree, temperature, side dish, and so on. He then coded the details of a particular order using a lettering system. For salad dressing, Blue

Cheese was always "B," Thousand Island always "T" and so on. Using this code with the other parts of the menu, he assigned the letters to an individual face and remembered the assignment. By creating a hierarchy of gist, he easily could apprehend the details.

J.C.'s strategy employs a principle well-known in the brain-science community: Memory is enhanced by creating associations between concepts. This experiment has been done hundreds of times, always achieving the same result: Words presented in a logically organized, hierarchical structure are much better remembered than words placed randomly—typically 40 percent better. This result baffles scientists to this day. Embedding associations between data points necessarily increases the number of items to be memorized. More pieces of intellectual baggage to inventory should make learning more difficult. But that is exactly not what was found. If we can derive the *meaning* of the words to one another, we can much more easily recall the details. Meaning *before* details.

John Bransford, a gifted education researcher who edited the well-received *How People Learn*, one day asked a simple question: In a given academic discipline, what separates novices from experts? Bransford eventually discovered six characteristics, one of which is relevant to our discussion: "[Experts'] knowledge is not simply a list of facts and formulas that are relevant to their domain; instead, their knowledge is organized around core concepts or 'big ideas' that guide their thinking about their domains."

Whether you are a waiter or a brain scientist, if you want to get the particulars correct, don't start with details. Start with the key ideas and, in a hierarchical fashion, form the details around these larger notions.

3) The brain cannot multitask

Multitasking, when it comes to paying attention, is a myth. The brain naturally focuses on concepts sequentially, one at a time. At first that might sound confusing; at one level the brain does multitask.

You can walk and talk at the same time. Your brain controls your heartbeat while you read a book. Pianists can play a piece with left hand and right hand simultaneously. Surely this is multitasking. But I am talking about the brain's ability to pay attention. It is the resource you forcibly deploy while trying to listen to a boring lecture at school. It is the activity that collapses as your brain wanders during a tedious presentation at work. This attentional ability is not capable of multitasking.

Recently, I agreed to help the high-school son of a friend of mine with some homework, and I don't think I will ever forget the experience. Eric had been working for about a half-hour on his laptop when I was ushered to his room. An iPod was dangling from his neck, the earbuds cranking out Tom Petty, Bob Dylan, and Green Day as his left hand reflexively tapped the backbeat. The laptop had at least 11 windows open, including two IM screens carrying simultaneous conversations with MySpace friends. Another window was busy downloading an image from Google. The window behind it had the results of some graphic he was altering for MySpace friend No. 2, and the one behind that held an old Pong game paused mid-ping.

Buried in the middle of this activity was a word-processing program holding the contents of the paper for which I was to provide assistance. "The music helps me concentrate," Eric declared, taking a call on his cell phone. "I normally do everything at school, but I'm stuck. Thanks for coming." Stuck indeed. Eric would make progress on a sentence or two, then tap out a MySpace message, then see if the download was finished, then return to his paper. Clearly, Eric wasn't concentrating on his paper. Sound like someone you know?

To put it bluntly, research shows that *we can't multitask*. We are biologically incapable of processing attention-rich inputs simultaneously. Eric and the rest of us must jump from one thing to the next.

To understand this remarkable conclusion, we must delve a little deeper into the third of Posner's trinity: the Executive Network. Let's look at what Eric's Executive Network is doing as he works on his paper and then gets interrupted by a "You've got mail!" prompt from his girlfriend, Emily.

STEP 1: SHIFT ALERT

To write the paper from a cold start, blood quickly rushes to the anterior prefrontal cortex in Eric's head. This area of the brain, part of the Executive Network, works just like a switchboard, alerting the brain that it's about to shift attention.

STEP 2: RULE ACTIVATION FOR TASK #1

Embedded in the alert is a two-part message, electricity sent crackling throughout Eric's brain. The first part is a search query to find the neurons capable of executing the paper-writing task. The second part encodes a command that will rouse the neurons, once discovered. This process is called "rule activation," and it takes several tenths of a second to accomplish. Eric begins to write his paper.

STEP 3: DISENGAGEMENT

While he's typing, Eric's sensory systems picks up the email alert from his girlfriend. Because the rules for writing a paper are different from the rules for writing to Emily, Eric's brain must disengage from the paper-writing rules before he can respond. This occurs. The switchboard is consulted, alerting the brain that another shift in attention is about to happen.

STEP 4: RULE ACTIVATION FOR TASK #2

Another two-part message seeking the rule-activation protocols for emailing Emily is now deployed. As before, the first is a command to find the writing-Emily rules, and the second is the activation command. Now Eric can pour his heart out to his sweetheart. As before, it takes several tenths of a second simply to perform the switch.

Incredibly, these four steps must occur in sequence *every time* Eric switches from one task to another. It is time-consuming. *And it is sequential.* That's why we can't multitask. That's why people find themselves losing track of previous progress and needing to "start over," perhaps muttering things like "Now where was I?" each time they switch tasks. The best you can say is that people who appear to be good at multitasking actually have good working memories, capable of paying attention to several inputs *one at a time.*

Here's why this matters: Studies show that a person who is interrupted takes 50 percent longer to accomplish a task. Not only that, he or she makes up to 50 percent more errors.

Some people, particularly younger people, are more adept at task-switching. If a person is familiar with the tasks, the completion time and errors are much less than if the tasks are unfamiliar. Still, taking your sequential brain into a multitasking environment can be like trying to put your right foot into your left shoe.

A good example is driving while talking on a cell phone. Until researchers started measuring the effects of cell-phone distractions under controlled conditions, nobody had any idea how profoundly they can impair a driver. It's like driving drunk. Recall that large fractions of a second are consumed every time the brain switches tasks. Cell-phone talkers are a half-second slower to hit the brakes in emergencies, slower to return to normal speed after an emergency, and more wild in their "following distance" behind the vehicle in front of them. In a half-second, a driver going 70 mph travels 51 feet. Given that 80 percent of crashes happen within three seconds of some kind of driver distraction, increasing your amount of task-switching increases your risk of an accident. More than 50 percent of the visual cues spotted by attentive drivers are missed by cell-phone talkers. Not surprisingly, they get in more wrecks than anyone except very drunk drivers.

It isn't just talking on a cell phone. It's putting on makeup, eating, rubber-necking at an accident. One study showed that simply reaching for an object while driving a car multiplies the risk of a crash or near-crash by nine times. Given what we know about the attention capacity of the human brain, these data are not surprising.

4) The brain needs a break

Our need for timed interruptions reminds me of a film called *Mondo Cane*, which holds the distinction of being the worst movie my parents reported ever seeing. Their sole reason for hating this movie was one disturbing scene: farmers force-feeding geese to make pâté de foie gras. Using fairly vigorous strokes with a pole, farmers literally stuffed food down the throats of these poor animals. When a goose wanted to regurgitate, a brass ring was fastened around its throat, trapping the food inside the digestive tract. Jammed over and over again, such nutrient oversupply eventually created a stuffed liver, pleasing to chefs around the world. Of course, it did nothing for the nourishment of the geese, who were sacrificed in the name of expediency.

My mother would often relate this story to me when she talked about being a good or bad teacher. "Most teachers overstuff their students," she would exclaim, "like those farmers in that awful movie!" When I went to college, I soon discovered what she meant. And now that I am a professor who has worked closely with the business community, I can see the habit close up. The most common communication mistakes? Relating too much information, with not enough time devoted to connecting the dots. Lots of force-feeding, very little digestion. This does nothing for the nourishment of the listeners, whose learning is often sacrificed in the name of expediency.

At one level, this is understandable. Most experts are so familiar with their topic that they forget what it is like to be a novice. Even if they remember, experts can become bored with having to repeat the fundamentals over and over again. In college, I found that a lot of my professors, because they had to communicate at such elementary levels, were truly fed up with teaching. They seemed to forget that the information was brand new to us, and that we needed the time to digest it, which meant a need for consistent breaks. How true indeed that expertise doesn't guarantee good teaching!

Such needs are not the case just in classrooms. I have observed similar mistakes in sermons, boardrooms, sales pitches, media stories—anywhere information from an expert needs to be transferred to a novice.

ideas

The 10-minute rule provides a way out of these problems. Here's the model I developed for giving a lecture, for which I was named the Hoechst Marion Rousell Teacher of the Year.

Lecture design: 10-minute segments

I decided that every lecture I'd ever give would come in discrete modules. Since the 10-minute rule had been known for many years, I decided the modules would last only 10 minutes. Each segment would cover a single core concept—always large, always general, always filled with "gist," *and always explainable in one minute*. Each class was 50 minutes, so I could easily burn through five large concepts in a single period. I would use the other 9 minutes in the segment to provide a detailed description of that single general concept. The trick was to ensure that each detail could be easily traced back to the general concept with minimal intellectual effort. I regularly took time out from content to explain the relationship between the detail and the core concept in clear and explicit terms. It was like allowing the geese to rest between stuffings.

Then came the hardest part: After 10 minutes had elapsed, I had to be finished with the core concept. Why did I construct it that way? Three reasons:

1) Given the tendency of an audience to check out 20 percent

of the way into a presentation, I knew I initially had only about 600 seconds to earn the right to be heard—or the next hour would be useless. I needed to do something after the 601st second to "buy" another 10 minutes.

2) The brain processes meaning before detail. Providing the gist, the core concept, *first* was like giving a thirsty person a tall glass of water. And the brain likes hierarchy. Starting with general concepts naturally leads to explaining information in a hierarchical fashion. You have to do the general idea *first*. And then you will see that 40 percent improvement in understanding.

3) It's key that the instructor explains the lecture plan at the beginning of the class, with liberal repetitions of "where we are" sprinkled throughout the hour. This prevents the audience from trying to multitask. If the instructor presents a concept without telling the audience where that concept fits into the rest of the presentation, the audience is forced to simultaneously listen to the instructor and attempt to divine where it fits into the rest of what the instructor is saying. This is the pedagogical equivalent of trying to drive while talking on a cell phone. Because it is impossible to pay attention to ANY two things at once, this will cause a series of millisecond delays throughout the presentation. The linkages must be clearly and repetitively explained.

Bait the hook

After 9 minutes and 59 seconds, the audience's attention is getting ready to plummet to near zero. If something isn't done quickly, the students will end up in successively losing bouts of an effort to stay with me. What do they need? Not more information of the same type. That would be like geese choking on the food with no real chance to digest. They also don't need some completely irrelevant cue that breaks them from their train of thought, making the information stream seem disjointed, unorganized, and patronizing. They need something so compelling that they blast through the 10-minute barrier and move on to new ground—something that triggers an orienting response toward the speaker and captures executive functions, allowing efficient learning.

Do we know anything so potentially compelling? We sure do. The ECS—emotionally competent stimuli. So, every 10 minutes in my lecture, I decided to give my audiences a break from the firehose of information and send them a relevant ECS, which I now call "hooks." As I did more teaching, I found the most successful hooks always followed these three principles:

1) The hook had to trigger an emotion. Fear, laughter, happiness, nostalgia, incredulity—the entire emotional palette could be stimulated, and all worked well. I deliberately employed Darwin here, describing some threatening event or, with appropriate taste, some reproductive event, even something triggering pattern matching. Narratives can be especially strong, especially if they are crisp and to the point.

2) The hook had to be relevant. It couldn't be just any story or anecdote. If I simply cracked a joke or delivered some irrelevant anecdote every 10 minutes, the presentation seemed disjointed. Or worse: The listeners began to mistrust my motives; they seemed to feel as if I were trying to entertain them at the expense of providing information. Audiences are really good at detecting disorganization, and they can become furious if they feel patronized. Happily, I found that if I made the hook very relevant to the provided content, the group moved from feeling entertained to feeling engaged. They stayed in the flow of my material, even though they were really taking a break.

3) The hook had to go between modules. I could place it at

the end of the 10 minutes, looking backward, summarizing the material, repeating some aspect of content. Or I could place it at the beginning of the module, looking forward, introducing new material, anticipating some aspect of content. I found that starting a lecture with a forward-looking hook relevant to the entire day's material was a great way to corral the attention of the class.

Exactly what did these hooks look like? This is where teaching can truly become imaginative. Because I work with psychiatric issues, case histories explaining some unusual mental pathology often rivet students to the upcoming (and drier) material. Business-related anecdotes can be fun, especially when addressing lay audiences in the corporate world. I often illustrate a talk about how brain science relates to business by addressing its central problem: vocabulary. I like the anecdote of the Electrolux Vacuum Cleaner company, a privately held corporation in Sweden trying to break into the North American market. They had plenty of English speakers on staff, but no Americans. Their lead marketing slogan? "If it sucks, it must be an Electrolux."

When I started placing hooks in my lectures, I immediately noticed changes in the audience members' attitudes. First, they were still interested at the end of the first 10 minutes. Second, they seemed able to maintain their attention for another 10 minutes or so, as long as another hook was supplied at the end. I could win the battle for their attention in 10-minute increments.

But then, halfway through the lecture, after I'd deployed two or three hooks, I found I could skip the fourth and fifth ones and still keep their attention fully engaged. I have found this to be true for students in 1994, when I first used the model, and in my lectures to this day.

Does that mean my model has harnessed the timing and power of emotional salience in human learning? That teachers and business professionals everywhere should drop whatever they are doing and incorporate its key features? I have no idea, but it would make sense to find out. The brain doesn't pay attention to boring things, and I am as sick of boring presentations as you are.

Do one thing at a time

The brain is a sequential processor, unable to pay attention to two things at the same time. Businesses and schools praise multitasking, but research clearly shows that it reduces productivity and increases mistakes. Try creating an interruption-free zone during the day—turn off your e-mail, phone, IM program, or BlackBerry—and see whether you get more done.

Summary

Rule #4

People don't pay attention to boring things.

* The brain's attentional "spotlight" can focus on only one thing at a time: no multitasking.

* We are better at seeing patterns and abstracting the meaning of an event than we are at recording detail.

* Emotional arousal helps the brain learn.

* Audiences check out after 10 minutes, but you can keep grabbing them back by telling narratives or creating events rich in emotion.

Get more at www.brainrules.net

short-term memory

Rule #5

Repeat to remember.

IT IS THE ULTIMATE intellectual flattery to be born with a mind so amazing that brain scientists voluntarily devote their careers to studying it. This impressive feat occurred with the owners of two such minds in the past century, and their remarkable brains provide much insight into human memory.

The first mind belongs to Kim Peek. He was born in 1951 with not one hint of his future intellectual greatness. He has an enlarged head, no corpus callosum, and a damaged cerebellum. He could not walk until age 4, and he can get catastrophically upset when he doesn't understand something, which is often. Diagnosing him in childhood as mentally disabled, his doctors wanted to place him in a mental institution. That didn't happen, mostly because of the nurturing efforts of Peek's father, who recognized that his son also had some very special intellectual gifts. One of those gifts is memory; Peek has one of the most prodigious ever recorded. He can read two pages at the same time, one with each eye, comprehending and remembering perfectly everything contained in the pages. Forever.

Though publicity shy, Peek's dad once granted writer Barry Morrow an interview with his son. It was conducted in a library, where Peek demonstrated to Morrow a familiarity with literally every book (and every author) in the building. He then started quoting ridiculous—and highly accurate—amounts of sports trivia. After a long discussion about the histories of certain United States wars (Revolutionary to Vietnam), Morrow felt he had enough. He decided right then and there to write a screenplay about this man. Which he did: the Oscar-winning film *Rain Man*.

What is going on in the uneven brain of Kim Peek? Does his mind belong in a cognitive freak show, or is it only an extreme example of normal human learning? Something very important is occurring in the first few moments his brain is exposed to information, and it's not so very different from what happens to the rest of us in the initial moments of learning.

The first few moments of learning give us the ability to remember something. The brain has different types of memory systems, many operating in a semi-autonomous fashion. We know so little about how they coordinate with each other that, to this date, memory is not considered a unitary phenomenon. We know the most about declarative memory, which involves something you can declare, such as "The sky is blue." This type of memory involves four steps: encoding, storage, retrieval, and forgetting. This chapter is about the first step. In fact, it is about the first few seconds of the first step. They are crucial in determining whether something that is initially perceived will also be remembered. Along the way, we will talk about our second famous mind. This brain, belonging to a man the research community called H.M., was legendary not for its extraordinary capabilities but for its extraordinary inabilities. We will also talk about the difference between bicycles and Social Security numbers.

memory and mumbo jumbo

Memory has been the subject of poets and philosophers for

centuries. At one level, memory is like an invading army, allowing past experiences to intrude continuously onto present life. That's fortunate. Our brains do not come fully assembled at birth, which means that most of what we know about the world has to be either experienced by us firsthand or taught to us secondhand. Our robust memory can provide great survival advantages—it is in large part why we've succeeded in overpopulating the planet. For a creature as physically weak as humans (compare your fingernail with the claw of even a simple cat, and weep with envy), not allowing experience to shape our brains would have meant almost certain death in the rough-and-tumble world of the open savannah.

But memory is more than a Darwinian chess piece. Most researchers agree that its broad influence on our brains is what truly makes us consciously aware. The names and faces of our loved ones, our own personal tastes, and especially our awareness of those names and faces and tastes, are maintained through memory. We don't go to sleep and then, upon awakening, have to spend a week relearning the entire world. Memory does this for us. Even the single most distinctive talent of human cognition, the ability to write and speak in a language, exists because of active remembering. Memory, it seems, makes us not only durable but also human.

Let's look at how it works. When researchers want to measure memory, they usually end up measuring retrieval. That's because in order to find out if somebody has committed something to memory, you have to ask if he or she can recall it. So, how do people recall things? Does the storage space carrying the record of some experience just sit there twiddling its thumbs in our brains, waiting for some command to trot out its contents? Can we investigate storage separately from retrieval? It has taken more than a hundred years of research just to get a glimmer of a definition of memory that makes sense to a scientist. The story began in the 19th century with a German researcher who performed the first real science-based inquiry into human memory. He did the whole thing with his own brain.

Hermann Ebbinghaus was born in 1850. As a young man, he looked like a cross between Santa Claus and John Lennon, with his bushy brown beard and round glasses. He is most famous for uncovering one of the most depressing facts in all of education: People usually forget 90 percent of what they learn in a class within 30 days. He further showed that the majority of this forgetting occurs within the first few hours after class. This has been robustly confirmed in modern times.

Ebbinghaus designed a series of experimental protocols with which a toddler might feel at ease: He made up lists of nonsense words, 2,300 of them. Each word consisted of three letters and a consonant-vowel-consonant construction, such as TAZ, LEF, REN, ZUG. He then spent the rest of his life trying to memorize lists of these words in varying combinations and of varying lengths.

With the tenacity of a Prussian infantryman (which, for a short time, he was), Ebbinghaus recorded for over 30 years his successes and failures. He uncovered many important things about human learning during this journey. He showed that memories have different life spans. Some memories hang around for only a few minutes, then vanish. Others persist for days or months, even for a lifetime. He also showed that one could increase the life span of a memory simply by repeating the information in timed intervals. The more repetition cycles a given memory experienced, the more likely it was to persist in his mind. We now know that the space between repetitions is the critical component for transforming temporary memories into more persistent forms. Spaced learning is greatly superior to massed learning.

Ebbinghaus's work was foundational. It was also incomplete. It did not, for example, separate the notion of memory from retrieval—the difference between learning something and recalling it later.

Go ahead and try to remember your Social Security number.

Easy enough? Your retrieval commands might include things like visualizing the last time you saw the card, or remembering the last time you wrote down the number. Now try to remember how to ride a bike. Easy enough? Hardly. You do not call up a protocol list detailing where you put your foot, how to create the correct angle for your back, where your thumbs are supposed to be. The contrast proves an interesting point: One does not recall how to ride a bike in the same way one recalls nine numbers in a certain order. The ability to ride a bike seems quite independent from any conscious recollection of the skill. You were consciously aware when you were remembering your Social Security number, but not when riding a bike. Do you need to have conscious awareness in order to experience a memory? Or is there more than one type of memory?

The answer seemed clearer as more data came in. The answer to the first question was no, which answered the second question. There are at least two types of memories: memories that involve conscious awareness and memories that don't. This awareness distinction gradually morphed into the idea that there were memories you could declare and there were memories you could not declare. Declarative memories are those that can be experienced in our conscious awareness, such as "this shirt is green," "Jupiter is a planet," or even a list of words. Nondeclarative memories are those that cannot be experienced in our conscious awareness, such as the motor skills necessary to ride a bike.

This does not explain everything about human memory. It does not even explain everything about declarative memory. But the rigor of Ebbinghaus gave future scientists their first real shot at mapping behavior onto a living brain. Then a 9-year-old boy was knocked off his bicycle, forever changing the way brain scientists thought about memory.

where memories go

In his accident, H.M. suffered a severe head injury that left him with epileptic seizures. These seizures got worse with age, eventually culminating in one major seizure and 10 blackout periods every seven days. By his late 20s, H.M. was essentially dysfunctional, of potential great harm to himself, in need of drastic medical intervention.

The desperate family turned to famed neurosurgeon William Scoville, who decided that the problem lay within the brain's temporal lobe (the brain region roughly located behind your ears). Scoville excised the inner surface of this lobe on both sides of the brain. This experimental surgery greatly helped the epilepsy. It also left H.M with a catastrophic memory loss. Since the day the surgery was completed, in 1953, H.M. has been unable to convert a new short-term memory into a long-term memory. He can meet you once and then an hour or two later meet you again, with absolutely no recall of the first visit.

He has lost the conversion ability Ebbinghaus so clearly described in his research more than 50 years before.

Even more dramatically, H.M. can no longer recognize his own face in the mirror. Why? As his face aged, some of his physical features changed. But, unlike the rest of us, H.M. cannot take this new information and convert it into a long-term form. This leaves him more or less permanently locked into a single idea about his appearance. When he looks in the mirror and does not see this single idea, he cannot identify to whom the image actually belongs.

As horrible as that is for H.M., it is of enormous value to the research community. Because researchers knew precisely what was taken from the brain, it was easy to map which brain regions controlled the Ebbinghaus behaviors. A great deal of credit for this work belongs to Brenda Milner, a psychologist who spent more than 40 years studying H.M. and laid the groundwork for much of our understanding about the nerves behind memory. Let's review for a moment the biology of the brain.

You recall the cortex—that wafer-thin layer of neural tissue that's about the size of a baby blanket when unfurled. It is composed of six discrete layers of cells. It's a busy place. Those cells process signals originating from many parts of the body, including those lassoed by your sense organs. They also help create stable memories, and that's where H.M.'s unfortunate experience becomes so valuable. Some of H.M.'s cortex was left perfectly intact; other regions, such as his temporal lobe, sustained heavy damage. It was a gruesome but ideal opportunity for studying how human memory forms.

This baby blanket doesn't just lay atop the brain, of course. As if the blanket were capable of growing complex, sticky root systems, the cortex adheres to the deeper structures of the brain by a hopelessly incomprehensible thicket of neural connections. One of the most important destinations of these connections is the hippocampus, which is parked near the center of your brain, one in each hemisphere. The hippocampus is specifically involved in converting short-term information into longer-term forms. As you might suspect, it is the very region H.M. lost during his surgery.

The anatomical relationship between the hippocampus and the cortex has helped 21st-century scientists further define the two types of memory. Declarative memory is any conscious memory system that is altered when the hippocampus and various surrounding regions become damaged. Non-declarative memory is defined as those unconscious memory systems that are NOT altered (or at least not greatly altered) when the hippocampus and surrounding regions are damaged. We're going to focus on declarative memory, a vital part of our everyday activities.

sliced and diced

Research shows that the life cycle of declarative memory can be divided into four sequential steps: encoding, storing, retrieving, and forgetting.

Encoding describes what happens at the initial moment of learning, that fleeting golden instant when the brain first encounters a new piece of declarative information. It also involves a whopping fallacy, one in which your brain is an active co-conspirator. Here's an example of this subversion, coming once again from the clinical observations of neurologist Oliver Sacks.

The case involves a low-functioning autistic boy named Tom, who has become quite famous for being able to "do" music (though little else). Tom never received formal instruction in music of any kind, but he learned to play the piano simply by listening to other people. Astonishingly, he could play complex pieces of music with the skill and artistry of accomplished professionals, on his first try after hearing the music exactly once. In fact, he has been observed playing the song "Fisher's Horn Pipe" with his left hand while simultaneously playing "Yankee Doodle Dandy" with his right hand while simultaneously singing "Dixie"! He also can play the piano backwards, that is, with his back to the keyboard and his hands inverted. Not bad for a boy who cannot even tie his own shoes.

When we hear about people like this, we are usually jealous. Tom absorbs music as if he could switch to the "on" position some neural recording device in his head. We think we also have this video recorder, only our model is not nearly as good. This is a common impression. Most people believe that the brain is a lot like a recording device—that learning is something akin to pushing the "record" button (and remembering is simply pushing "playback"). Wrong. In the real world of the brain—Tom's or yours—nothing could be further from the truth. The moment of learning, of encoding, is so mysterious and complex that we have no metaphor to describe what happens to our brains in those first fleeting seconds.

The little we do know suggests it is like a blender left running with the lid off. The information is literally sliced into discrete pieces as it enters the brain and splattered all over the insides of our mind. Stated formally, signals from different sensory sources are registered in separate brain areas. The information is fragmented and redistributed the instant the information is encountered. If you look at a complex picture, for example, your brain immediately extracts the diagonal lines from the vertical lines and stores them in separate areas. Same with color. If the picture is moving, the fact of its motion will be extracted and stored in a place separate than if the picture were static.

This separation is so violent, and so pervasive, it even shows up when we perceive exclusively human-made information, such as parts of a language. One woman suffered a stroke in a specific region of her brain and lost the ability to use written vowels. You could ask her to write down a simple sentence, such as "Your dog chased the cat," and it would look like this:

Y _ _ r d _ g ch _ s _ d t h _ c _ t.

There would be a place for every letter, but the vowels' spots were left blank! So we know that vowels and consonants are not stored in the same place. Her stroke damaged some kind of connecting wiring. That is exactly the opposite of the strategy a video recorder uses to record things. If you look closely, however, the blender effect goes much deeper. Even though she lost the ability to fill in the vowels of a given word, she has perfectly preserved the place where the vowel should go. Using the same logic, it appears that the place where a vowel should go is stored in a separate area from the vowel itself: Content is stored separately from its context/container.

Hard to believe, isn't it? The world appears to you as a unified whole. If the interior brain function tells us that it is not, how then do we keep track of everything? How do features that are registered separately, including the vowels and consonants in this sentence, become reunited to produce perceptions of continuity? It is a question that has bothered researchers for years and has been given its own special name. It is called the "binding problem," from the idea that certain thoughts are bound together in the brain to provide continuity. We have no idea how the brain routinely and effortlessly gives us this illusion of stability.

Not that there aren't hints. Close inspection of the initial moments of learning, the encoding stage, has supplied insights into not only the binding problem, but human learning of any kind. It is to these hints that we now turn.

automatic or stick shift?

To encode information means to convert data into, well, a code. Creating codes always involves translating information from one form into another, usually for transmission purposes, often to keep something secret. From a physiological point of view, encoding is the conversion of external sources of energy into electrical patterns the brain can understand. From a purely psychological point of view, it is the manner in which we apprehend, pay attention to, and ultimately organize information for storage purposes. Encoding, from both perspectives, prepares information for further processing. It is one of the many intellectual processes the Rain Man, Kim Peek, is so darn good at.

The brain is capable of performing several types of encoding. One type of encoding is automatic, which can be illustrated by talking about what you had for dinner last night, or The Beatles. The two came together for me on the evening of an amazing Paul McCartney concert I attended a few years ago. If you were to ask me what I had for dinner before the concert and what happened on stage, I could tell you about both events in great detail. Though the actual memory is very complex (composed of spatial locations, sequences of events, sights, smells, tastes, etc.), I did not have to write down some exhaustive list of its varied experiences, then try to remember the list in detail just in case you asked me about my evening. This is because my brain deployed a certain type of encoding scientists call automatic processing. It is the kind occurring with glorious unintentionality, requiring minimal attentional effort. It is very easy to recall data that have been encoded via this process. The memories seem bound all together into a cohesive, readily retrievable form.

Automatic processing has an evil twin that isn't nearly so accommodating, however. As soon as the Paul McCartney tickets went on sale, I dashed to the purchasing website, which required my password for entrance. And I couldn't remember my password! Finally, I found the right one and snagged some good seats. But trying to commit these passwords to memory is quite a chore, and I have a dozen or so passwords written on countless lists, scattered throughout my house. This kind of encoding, initiated deliberately, requiring conscious, energy-burning attention, is called effortful processing. The information does not seem bound together well at all, and it requires a lot of repetition before it can be retrieved with the ease of automatic processing.

encoding test

There are still other types of encoding, three of which can be illustrated by taking the quick test below. Examine the capitalized word beside the number, then answer the question below it.

1) FOOTBALL
Does this word fit into the sentence "I turned around to fight _____"?

2) LEVEL
Does this word rhyme with evil?

3) MINIMUM
Are there any circles in these letters?

Answering each question requires very different intellectual skills, which researchers now know underlie different types of encoding. The first sentence illustrates what is called semantic encoding. Answering the question properly means paying attention to the definitions of words. The second sentence illustrates a process called phonemic encoding, involving a comparison between the sounds of words. The third is called structural encoding. It is the most superficial type, and it simply asks for a visual inspection of shapes. The type of encoding you perform on a given piece of information as it enters your head has a great deal to do with your ability to remember the information at a later date.

the electric slide

Encoding also involves transforming any outside stimulus into the electrical language of the brain, a form of energy transfer. All types of encoding initially follow the same pathway, and generally the same rules. For example, the night of Sir Paul's concert, I stayed with a friend who owned a beautiful lake cabin inhabited by a very large and hairy dog. Late next morning, I decided to go out and play fetch with this friendly animal. I made the mistake of throwing the stick into the lake and, not owning a dog in those days, had no idea what was about to happen to me when the dog emerged.

Like some friendly sea monster from Disney, the dog leapt from the water, ran at me full speed, suddenly stopped, then started to shake violently. With no real sense that I should have moved, I got sopping wet.

What was occurring in my brain in those moments? As you know, the cortex quickly is consulted when a piece of external information invades our brains—in this case, a slobbery, soaking wet Labrador. I see the dog coming out of the lake, which really means I see patterns of photons bouncing off the Labrador. The instant those photons hit the back of my eyes, my brain converts them into patterns of electrical activity and routes the signals to the back of my head (the visual cortex in the occipital lobe). Now my brain can see the dog. In the initial moments of this learning, I have transformed the energy of light into an electrical language the brain fully understands. Beholding this action required the coordinated activation of thousands of cortical regions dedicated to visual processing.

The same is also true of other energy sources. My ears pick up the sound waves of the dog's loud bark, and I convert them into the same brain-friendly electrical language to which the photons patterns were converted. These electrical signals will also be routed to the cortex, but to the auditory cortex instead of the visual cortex. From a nerve's perspective, those two centers are a million miles away from each other. This conversion and this vastly individual routing are true of all the energy sources coming into my brain, from the feel of the sun on my skin to the instant I unexpectedly and unhappily got soaked by the dog shaking off lake water. Encoding involves all of our senses, and their processing centers are scattered throughout the brain.

This is the heart of the blender. In one 10-second encounter with an overly friendly dog, my brain recruited hundreds of different brain regions and coordinated the electrical activity of millions of neurons. My brain was recording a single episode, and doing so over vast neural differences, all in about the time it takes to blink your eyes.

Years have passed since I saw Sir Paul and got drenched by the dog. How do we keep track of it all? And how do we manage to manage these individual pieces for years? This binding problem, a phenomenon that keeps tabs on farflung pieces of information, is a great question with, unfortunately, a lousy answer. We really don't know how the brain keeps track of things. We have given a name to the total number of changes in the brain that first encode information (where we have a record of that information). We call it an engram. But we might as well call them donkeys for all we understand about them.

The only insight we have into the binding problem comes from studying the encoding abilities of a person suffering from Balint's Syndrome. This disorder occurs in people who have damaged both sides of their parietal cortex. The hallmark of people with Balint's Syndrome is that they are functionally blind. Well, sort of. They can see objects in their visual field, but only one at a time (a symptom called simultanagnosia). Funny thing is, if you ask them where the single object is, they respond with a blank stare. Even though they can see it, they cannot tell you where it is. Nor can they tell you if the object is moving toward them or away from them. They have no external spatial frame of reference upon which to place the objects they see, no way to bind the image to other features of the input. They've lost explicit spatial awareness, a trait needed in any type of binding exercise. That's about as close as anyone has ever come to describing the binding problem at the neurological level. This tells us very little about how the brain solves the problem, of course. It only tells us about some of the areas involved in the process.

cracking the code

Despite their wide reach, scientists have found that all encoding processes have common characteristics. Three of these hold true promise for real-world applications in both business and education.

1) The more elaborately we encode information at the moment of learning, the stronger the memory.

When encoding is elaborate and deep, the memory that forms is much more robust than when encoding is partial and cursory. This can be demonstrated in an experiment you can do right now with any two groups of friends. Have them gaze at the list of words below for a few minutes.

Tractor	Pastel	Airplane
Green	Quickly	Jump
Apple	Ocean	Laugh
Zero	Nicely	Tall
Weather	Countertop	

Tell Group #1 to determine the number of letters that have diagonal lines in them and the number that do not. Tell Group #2 to think about the meaning of each word and rate, on a scale of 1 to 10, how much they like or dislike the word. Take the list away, let a few minutes pass, and then ask each group to write down as many words as possible. The dramatic results you get have been replicated in laboratories around the world. The group that processes the meaning of the words always remembers two to three times as many words as the group that looked only at the architecture of the individual letters. We did a form of this experiment when we discussed levels of encoding and I asked you about the number of circles in the word … remember what it was? You can do a similar experiment using pictures. You can even do it with music. No matter the sensory input, the results are always the same.

At this point, you might be saying to yourself, "Well, duh!" Isn't it obvious that the more meaning something has, the more memorable it becomes? Most researchers would answer, "Well, yeah!" The very naturalness of the tendency proves the point. Hunting for diagonal lines in the word "apple" is not nearly as elaborate as remembering wonderful Aunt Mabel's apple pie, then rating the pie, and thus the word, a "10." We remember things much better the more elaborately we encode what we encounter, especially if we can personalize it. The trick for business professionals, and for educators, is to present bodies of information so compelling that the audience does this on their own, spontaneously engaging in deep and elaborate encoding.

It's a bit weird if you think about it. Making something more elaborate usually means making it more complicated, which should be more taxing to a memory system. But it's a fact: More complexity means greater learning.

2) A memory trace appears to be stored in the same parts of the brain that perceived and processed the initial input.

This idea is so counterintuitive that it may take an urban legend to explain it. At least, I think it's an urban legend, coming from the mouth of the keynote speaker at a university administrators' luncheon I once attended. He told the story of the wiliest college president he ever encountered. The institute had completely redone its grounds in the summer, resplendent with fountains and beautifully manicured lawns. All that was needed was to install the sidewalks and walkways where the students could access the buildings. But there was no design for these paths. The construction workers were anxious to install them and wanted to know what the design would be, but the wily president refused to give any. He frowned. "These asphalt paths will be permanent. Install them next year, please. I will give you the plans then." Disgruntled but compliant, the construction workers waited.

The school year began, and the students were forced to walk on the grass to get to their classes. Very soon, defined trails started appearing all over campus, as well as large islands of beautiful green lawn. By the end of the year, the buildings were connected by paths in a surprisingly efficient manner. "Now," said the president to the contractors who had waited all year, "you can install the permanent sidewalks and pathways. But you need no design. Simply fill in all the paths you see before you!" The initial design, created by the initial input, also became the permanent path.

The brain has a storage strategy remarkably similar to the wily president's plan. The neural pathways initially recruited to process new information end up becoming the permanent pathways the brain reuses to store the information. New information penetrating into the brain can be likened to the students initially creating the dirt paths across a pristine lawn. The final storage area can be likened to the time those pathways were permanently filled with asphalt. They are the same pathways, and that's the point.

What does this mean for the brain? The neurons in the cortex are active responders in any learning event, and they are deeply involved in permanent memory storage. This means the brain has no central happy hunting ground where memories go to be infinitely retrieved. Instead, memories are distributed all over the surface of the cortex. This may at first seem hard to grasp. Many people would like the brain to act like a computer, complete with input detectors (like a keyboard) connected to a central storage device. Yet the data suggest that the human brain has no hard drive separate from its initial input detectors. That does not mean memory storage is spread evenly across the brain's neural landscape. Many brain regions are involved in representing even single inputs, and each region contributes something different to the entire memory. Storage is a cooperative event.

3) Retrieval may best be improved by replicating the conditions surrounding the initial encoding.

In one of the most unusual experiments performed in cognitive psychology, the brain function of people standing around on dry ground in wet suits was compared with the brain function of people floating in about 10 feet of water, also in wet suits. Both groups of deep-sea divers listened to somebody speak 40 random words. The divers were then tested for their ability to recall the list of words. The group that heard the words while in the water got a 15 percent better score if they were asked to recall the words while back in those same 10 feet than if they were on the beach. The group that heard the words on the beach got a 15 percent better score if they were asked to recall the words while suited on the beach than if in 10 feet of water. It appeared that memory worked best if the environmental conditions at retrieval mimicked the environmental conditions at encoding. Is it possible that the second characteristic, which tries to store events using the same neurons recruited initially to encode events, is in operation in this third characteristic?

The tendency is so robust that memory is even improved under conditions where learning of any kind should be crippled. These experiments have been done incorporating marijuana and even laughing gas (nitrous oxide). This third characteristic even responds to mood. Learn something while you are sad and you will be able to recall it better if, at retrieval, you are somehow suddenly made sad. The condition is called context-dependent or state-dependent learning.

ideas

We know that information is remembered best when it is elaborate, meaningful, and contextual. The quality of the encoding stage—those earliest moments of learning—is one of the single greatest predictors of later learning success. What can we do to take advantage of that in the real world?

First, we can take a lesson from a shoe store I used to visit as a little boy. This shoe store had a door with three handles at different heights: one near the very top, one near the very bottom, and one in the middle. The logic was simple: The more handles on the door, the more access points were available for entrance, regardless of the strength or age of customer. It was a relief for a 5-year-old—a door I could actually reach! I was so intrigued with the door that I used to dream about it. In my dreams, however, there were hundreds of handles, all capable of opening the door to this shoe store.

"Quality of encoding" really means the number of door handles one can put on the entrance to a piece of information. The more handles one creates at the moment of learning, the more likely the information is to be accessed at a later date. The handles we can add revolve around content, timing, and environment.

Real-world examples

The more a learner focuses on the meaning of the presented information, the more elaborately the encoding is processed. This principle is so obvious that it is easy to miss. What it means is this: When you are trying to drive a piece of information into your brain's memory systems, make sure you understand exactly what that information means. If you are trying to drive information into someone else's brain, make sure they know what it means.

The directive has a negative corollary. If you don't know what the learning means, don't try to memorize the information by rote and pray the meaning will somehow reveal itself. And don't expect your students will do this either, especially if you have done an inadequate job of explaining things. This is like looking at the number of diagonal lines in a word and attempting to use this strategy to remember the words.

How does one communicate meaning in such a fashion that learning is improved? A simple trick involves the liberal use of relevant real-world examples embedded in the information, constantly peppering main learning points with meaningful experiences. This can be done by the learner studying after class or, better, by the teacher during the actual learning experience. This has been shown to work in numerous studies.

In one experiment, groups of students read a 32-paragraph paper about a fictitious foreign country. The introductory paragraphs in the paper were highly structured. They contained either no examples, one example, or two or three consecutive examples of the main theme that followed. The results were clear: The greater the number of examples in the paragraph, the more likely the information was to be remembered. It's best to use real-world situations familiar to the learner. Remember wonderful Aunt Mabel's apple pie? This wasn't an abstract food cooked by a stranger; it was real food cooked by a loving relative. The more personal an example, the more richly it becomes encoded and the more readily it is remembered.

Why do examples work? They appear to take advantage of the brain's natural predilection for pattern matching. Information is more readily processed if it can be immediately associated with information already present in the learner's brain. We compare the two inputs, looking for similarities and differences as we encode the new information. Providing examples is the cognitive equivalent of adding more handles to the door. Providing examples makes the information more elaborative, more complex, better encoded, and therefore better learned.

Compelling introductions

Introductions are everything. As an undergraduate, I had a professor who can thoughtfully be described as a lunatic. He taught a class on the history of cinema, and one day he decided to illustrate for us how art films traditionally depict emotional vulnerability. As he went through the lecture, he literally began taking off his clothes. He first took off his sweater and then, one button at a time, began removing his shirt, down to his T-shirt. He unzipped his trousers, and they fell around his feet, revealing, thank goodness, gym clothes. His eyes were shining as he exclaimed, "You will probably never forget now that some films use physical nudity to express emotional vulnerability. What could be more vulnerable than being naked?" We were thankful that he gave us no further details of his example.

I will never forget the introduction to this unit in my film class, though I hardly recommend imitating his example on a regular basis. But its memorability illustrates the timing principle: If you are a student, whether in business or education, the events that happen the first time you are exposed to a given information stream play a disproportionately greater role in your ability to accurately retrieve it at a later date. If you are trying to get information across to someone, your ability to create a compelling introduction may be the most important single factor in the later success of your mission.

Why this emphasis on the initial moments? Because the memory of an event is stored in the same places that were initially recruited to perceive the learning event. The more brain structures recruited—the more door handles created—at the moment the learning, the easier it is to gain access to the information.

Other professions have stumbled onto this notion. Budding directors are told by their film instructors that the audience needs to be hooked in the first 3 minutes after the opening credits to make the film compelling (and financially successful). Public speaking professionals say that you win or lose the battle to hold your audience in the first 30 seconds of a given presentation.

What does that mean for business professionals attempting to create a compelling presentation? Or educators attempting to introduce a complex new topic? Given the importance of the findings to the success of these professions, you might expect that some rigorous scientific literature exists on this topic. Surprisingly, very little data exist about how brains pay attention to issues in real-world settings, as we discussed in the Attention chapter. The data that do exist suggest that film instructors and public speakers are on to something.

Familiar settings

We know the importance of learning and retrieval taking place under the same conditions, but we don't have a solid definition of "same conditions." There are many ways to explore this idea.

I once gave a group of teachers advice about how to counsel parents who wanted to teach both English and Spanish at home. One dissatisfying finding is that for many kids with this double exposure, language acquisition rates for both go down, sometimes considerably. I recounted the data about the underwater experiments and then suggested that the families create a "Spanish Room." This would be a room with a rule: Only the Spanish language could be spoken in it. The room could be filled with Hispanic artifacts, with large pictures of Spanish words. All Spanish would be taught there, and no English. Anecdotally, the parents have told me that it works.

This way, the encoding environments and retrieving environments could be equivalent. At the moment of learning, many environmental features—even ones irrelevant to the learning goals—may become encoded into the memory right along with the goals. Environment makes the encoding more elaborate, the equivalent of putting more handles on the door. When these same environmental cues are encountered, they may lead directly to the learning goals simply because they were embedded in the original trace.

American marketing professionals have known about this

phenomenon for years. What if I wrote the words "wind-up pink bunny," "pounding drum," and "going-and-going," then told you to write another word or phrase congruent with those previous three? No formal relationship exists between any of these words, yet if you lived in the United States for a long period of time, most of you probably would write words such as "battery" or "Energizer." Enough said.

What does it mean to make encoding and retrieving environments equivalent in the real world of business and education? The most robust findings occur when the environments exist in dramatically different contexts from the norm (underwater vs. on a beach is about as dramatic as it gets). But how different from normal life does the setup need to be to obtain the effect?

It could be as simple as making sure that an oral examination is studied for orally, rather than by reviewing written material. Or perhaps future airplane mechanics should be taught about engine repair in the actual shop where the repairs will occur.

Summary

Rule #5
Repeat to remember.

* The brain has many types of memory systems. One type follows four stages of processing: encoding, storing, retrieving, and forgetting.

* Information coming into your brain is immediately split into fragments that are sent to different regions of the cortex for storage.

* Most of the events that predict whether something learned *also will be remembered* occur in the first few seconds of learning. The more elaborately we encode a memory during its initial moments, the stronger it will be.

* You can improve your chances of remembering something if you reproduce the environment in which you first put it into your brain.

Get more at www.brainrules.net

long-term memory

Rule #6

Remember to repeat.

FOR MANY YEARS, TEXTBOOKS described the birth of a memory using a metaphor involving cranky dockworkers, a large bookstore, and a small loading dock. An event to be processed into memory was likened to somebody dropping off a load of books onto the dock. If a dockworker hauled the load into the vast bookstore, it became stored for a lifetime. Because the loading dock was small, only a few loads could be processed at any one time. If someone dumped a new load of books on the dock before the previous ones were removed, the cranky workers simply pushed the old ones over the side.

Nobody uses this metaphor anymore, and there are ample reasons to wish it good riddance. Short-term memory is a much more active, much less sequential, far more complex process than the metaphor suggests. We now suspect that short-term memory is actually a collection of temporary memory capacities. Each capacity specializes in processing a specific type of information. Each operates in a parallel fashion with the others. To reflect this multifaceted talent, short-term memory is now called working memory. Perhaps the best way to explain working memory is to describe it in action.

I can think of no better illustration than the professional chess world's first real rock star: Miguel Najdorf. Rarely was a man more at ease with his greatness than Najdorf. He was a short, dapper fellow gifted with a truly enormous voice, and he had an annoying tendency to poll members of his audience on how they thought he was doing. Najdorf in 1939 traveled to a competition in Buenos Aires with the national team. Two weeks later, Germany invaded Najdorf's home country of Poland. Unable to return, Najdorf rode out the Holocaust tucked safely inside Argentina. He lost his parents, four brothers, and his wife to the concentration camps. In hopes that any remaining family might read about it and contact him, he once played 45 games of chess simultaneously, as a publicity stunt. He won 39 of these games, drew 4, and lost 2. While that is amazing in its own right, the truly phenomenal part is that he played all 45 games in all 11 hours *blindfolded*.

You did not read that wrong. Najdorf never physically saw any of the chessboards or pieces; he played each game in his mind. From the verbal information he received with each move, to his visualizations of each board, several components of working memory were working simultaneously in Najdorf's mind. This allowed him to function in his profession, just as they do in yours and mine (though perhaps with a slightly different efficiency).

Working memory is now known to be a busy, temporary workspace, a desktop the brain uses to process newly acquired information. The man most associated with characterizing it is Alan Baddeley, a British scientist who looks unnervingly like the angel Clarence Oddbody in the movie *It's a Wonderful Life*. Baddeley is most famous for describing working memory as a three-component model: auditory, visual, and executive.

The first component allows us to retain some auditory information, and it is assigned to information that is linguistic.

Baddeley called it a phonological loop. Najdorf was able to use this component because his opponents were forced to declare their moves verbally.

The second component allows us to retain some visual information; this memory register is assigned to any images and spatial input the brain encounters. Baddeley called it a visuo-spatial sketchpad. Najdorf would have used it as he visualized each game.

The third component is a controlling function called the central executive, which keeps track of all activities throughout working memory. Najdorf used this ability to separate one game from another.

In later publications, Baddeley proposed a fourth component, called the episodic buffer, assigned to any stories a person might hear. This buffer has not been investigated extensively. Regardless of the number of parallel systems ultimately discovered, researchers agree that all share two important characteristics: All have a limited capacity, and all have a limited duration. If the information is not transformed into a more durable form, it will soon disappear. As you recall, our friend Ebbinghaus was the first to demonstrate the existence of two types of memory systems, a short form and a long form. He further demonstrated that repetition could convert one into the other under certain conditions. The process of converting short-term memory traces to longer, sturdier forms is called consolidation.

consolidation

At first, a memory trace is flexible, labile, subject to amendment, and at great risk for extinction. Most of the inputs we encounter in a given day fall into this category. But some memories stick with us. Initially fragile, these memories strengthen with time and become remarkably persistent. They eventually reach a state where they appear to be infinitely retrievable and resistant to amendment. As we shall see, however, they may not be as stable as we think. Nonetheless, we call these forms long-term memories.

Like working memory, there appear to be different forms of

long-term memory, most of which interact with one another. Unlike working memory, however, there is not as much agreement as to what those forms are. Most researchers believe there are semantic memory systems, which tend to remember things like your Aunt Martha's favorite dress or your weight in high school. Most also believe there is episodic memory, in charge of remembering "episodes" of past experiences, complete with characters, plots, and time stamps (like your 25th high school reunion). One of its subsets is autobiographical memory, which features a familiar protagonist: you. We used to think that consolidation, the mechanism that guides this transformation into stability, affected only newly acquired memories. Once the memory hardened, it never returned to its initial fragile condition. We don't think that anymore.

Consider the following story, which happened while I was watching a TV documentary with my then 6-year-old son. It was about dog shows. When the camera focused on a German shepherd with a black muzzle, an event that occurred when *I* was about his age came flooding back to my awareness.

In 1960, our backyard neighbor owned a dog he neglected to feed every Saturday. In response to hunger cues, the dog bounded over our fence precisely at 8 a.m. every Saturday, ran toward our metal garbage cans, tipped out the contents, and began a morning repast. My dad got sick of this dog and decided one Friday night to electrify the can in such fashion that the dog would get shocked if his wet little nose so much as brushed against it. Next morning, my dad awakened our entire family early to observe his "hot dog" show. To Dad's disappointment, the dog didn't jump over the fence until about 8:30, and he didn't come to eat. Instead he came to mark his territory, which he did at several points around our backyard. As the dog moved closer to the can, my Dad started to smile, and when the dog lifted his leg to mark our garbage can, my Dad exclaimed, "Yes!" You don't have to know the concentration of electrolytes in mammalian urine to know that when the dog marked his territory on our garbage can, he also completed a mighty circuit. His cranial neurons ablaze, his reproductive future suddenly in serious question, the dog howled, bounding back to his owner. The dog never set foot in our backyard again; in fact, he never came within 100 yards of our house. Our neighbor's dog was a German shepherd with a distinct black muzzle, just like the one in the television show I was now watching. I had not thought of the incident in years.

What physically happened to my dog memory when summoned back to awareness? There is increasing evidence that when previously consolidated memories are recalled from long-term storage into consciousness, they revert to their previously labile, unstable natures. Acting as if newly minted into working memory, these memories may need to become reprocessed if they are to remain in a durable form. That means the hot dog story is forced to restart the consolidation process all over again, *every time it is retrieved*. This process is formally termed reconsolidation. These data have a number of scientists questioning the entire notion of stability in human memory. If consolidation is not a sequential one-time event but one that occurs repeatedly every time a memory trace is reactivated, it means permanent storage exists in our brains only for those memories we choose not to recall! Oh, good grief. Does this mean that we can never be aware of something permanent in our lives? Some scientists think this is so. And if it is true, the case I am about to make for repetition in learning is ridiculously important.

retrieval

Like many radical university professors, our retrieval systems are powerful enough to alter our conceptions of the past while offering nothing substantial to replace them. Exactly how that happens is an important but missing piece of our puzzle. Still, researchers have organized the mechanisms of retrieval into two general models. One passively imagines libraries. The other aggressively imagines crime scenes.

In the library model, memories are stored in our heads the same way books are stored in a library. Retrieval begins with a command to browse through the stacks and select a specific volume. Once selected, the contents of the volume are brought into conscious awareness, and the memory is retrieved. This tame process is sometimes called reproductive retrieval.

The other model imagines our memories to be more like a large collection of crime scenes, complete with their own Sherlock Holmes. Retrieval begins by summoning the detective to a particular crime scene, a scene which invariably consists of a fragmentary memory. Upon arrival, Mr. Holmes examines the partial evidence available. Based on inference and guesswork, the detective then invents a reconstruction of what was actually stored. In this model, retrieval is not the passive examination of a fully reproduced, vividly detailed book. Rather, retrieval is an active investigative effort to re-create the facts based on fragmented data.

Which is correct? The surprising answer is both. Ancient philosophers and modern scientists agree that we have different types of retrieval systems. Which one we use may depend upon the type of information being sought, and how much time has passed since the initial memory was formed. This unusual fact requires some explanation.

mind the gap

At relatively early periods post-learning (say minutes to hours to days), retrieval systems allow us to reproduce a fairly specific and detailed account of a given memory. This might be likened to the library model. But as time goes by, we switch to a style more reminiscent of the Sherlock Holmes model. The reason is that the passage of time inexorably leads to a weakening of events and facts that were once clear and chock-full of specifics. In an attempt to fill in missing gaps, the brain is forced to rely on partial fragments, inferences, outright guesswork, and often (most disturbingly) other memories not related to the actual event. It is truly reconstructive in nature, much like a detective with a slippery imagination. This is all in the service of creating a coherent story, which, reality notwithstanding, brains like to do. So, over time, the brain's many retrieval systems seem to undergo a gradual switch from specific and detailed reproductions to this more general and abstracted recall.

Pretend you are a freshman in high school and know a psychiatrist named Daniel Offer. Taking out a questionnaire, Dr. Dan asks you to answer some questions that are really none of his business: Was religion helpful to you growing up? Did you receive physical punishment as discipline? Did your parents encourage you to be active in sports? And so on. Now pretend it is 34 years later. Dr. Dan tracks you down, gives you the same questionnaire, and asks you to fill it out. Unbeknownst to you, he still has the answers you gave in high school, and he is out to compare your answers. How well do you do? In a word, horribly. In fact, the memories you encoded as adolescents bear very little resemblance to the ones you remember as adults, as Dr. Dan, who had the patience to actually do this experiment, found out. Take the physical punishment question. Though only a third of adults recalled any physical punishment, such as spanking, Dr. Dan found that almost 90 percent of the adolescents had answered the question in the affirmative. These data are only some that demonstrate the powerful inaccuracy of the Sherlock Holmes style of retrieval.

This idea that the brain might cheerily insert false information to make a coherent story underscores its admirable desire to create organization out of a bewildering and confusing world. The brain constantly receives new inputs and needs to store some of them in the same head already occupied by previous experiences. It makes sense of its world by trying to connect new information to previously encountered information, which means that new information routinely resculpts previously existing representations and sends the re-created whole back for new storage. What does this mean?

Merely that present knowledge can bleed into past memories and become intertwined with them as if they were encountered together. Does that give you only an approximate view of reality? You bet it does. This tendency, by the way, can drive the criminal-justice system crazy.

repetition

Given this predilection for generalizing, is there any hope of creating reliable long-term memories? As the Brain Rule cheerily suggests, the answer is yes. Memory may not be fixed at the moment of learning, but repetition, doled out in specifically timed intervals, is the fixative. Given its potential relevance to business and education, it is high time we talked about it.

Here's a test that involves the phonological loop of working memory. Gaze at the following list of characters for about 30 seconds, then cover it up before you read the next paragraph.

3 $ 8 ? A % 9

Can you recall the characters in the list without looking at them? Were you able to do this without internally rehearsing them? Don't be alarmed if you can't. The typical human brain can hold about seven pieces of information for less than 30 seconds! If something does not happen in that short stretch of time, the information becomes lost. If you want to extend the 30 seconds to, say, a few minutes, or even an hour or two, you will need to consistently re-expose yourself to the information. This type of repetition is sometimes called maintenance rehearsal. We now know that maintenance rehearsal is mostly good for keeping things in working memory—that is, for a short period of time. We also know there is a better way to push information into long-term memory. To describe it, I would like to relate the first time I ever saw somebody die.

Actually, I saw eight people die. Son of a career Air Force official,

I was very used to seeing military airplanes in the sky. But I looked up one afternoon to see a cargo plane do something I had never seen before or since. It was falling from the sky, locked in a dead man's spiral. It hit the ground maybe 500 feet from where I stood, and I felt both the shock wave and the heat of the explosion.

There are two things I could have done with this information. I could have kept it entirely to myself, or I could have told the world. I chose the latter. After immediately rushing home to tell my parents, I called some of my friends. We met for sodas and began talking about what had just happened. The sounds of the engine cutting out. Our surprise. Our fear. As horrible as the accident was, we talked about it so much in the next week that the subject got tiresome. One of my teachers actually forbade us from bringing it up during class time, threatening to make T-shirts saying, "You've done enough talking."

Why do I still remember the details of this story? T-shirt threat notwithstanding, my eagerness to yap about the experience provided the key ingredient. The gabfest after the accident forced a consistent re-exposure to the basic facts, followed by a detailed elaboration of our impressions. The phenomenon is called elaborative rehearsal, and it's the type of repetition shown to be most effective for the most robust retrieval. A great deal of research shows that thinking or talking about an event *immediately after it has occurred* enhances memory for that event, even when accounting for differences in type of memory. This tendency is of enormous importance to law-enforcement professionals. It is one of the reasons why it is so critical to have a witness recall information as soon as is humanely possible after a crime.

Ebbinghaus showed the power of repetition in exhaustive detail almost 100 years ago. He even created "forgetting curves," which showed that a great deal of memory loss occurs in the first hour or two after initial exposure. He demonstrated that this loss could be lessened by deliberate repetitions. This notion of timing in the midst of re-exposure is so critical, I am going to explore it in three ways.

space out the input

Much like concrete, memory takes an almost ridiculous amount of time to settle into its permanent form. And while it is busy hardening, human memory is maddeningly subject to amendment. This probably occurs because newly encoded information can reshape and wear away previously existing traces. Such interference is especially true when learning is supplied in consecutive, uninterrupted glops, much like what happens in most boardrooms and schoolrooms. The probability of confusion is increased when content is delivered in unstoppable, unrepeated waves, poured into students as if they were wooden forms.

But there is happy news. Such interference does not occur if the information is delivered in deliberately spaced repetition cycles. Indeed, repeated exposure to information in specifically timed intervals provides the most powerful way to fix memory into the brain. Why does this occur? When the electrical representations of information to be learned are built up slowly over many repetitions, the neural networks recruited for storage gradually remodel the overall representation *and do not interfere* with networks previously recruited to store similarly learned information. This idea suggests that continuous repetition cycles create experiences capable of *adding* to the knowledge base, rather than interfering with the resident tenants.

There is an area of the brain that always becomes active when a vivid memory is being retrieved. The area is within the left inferior prefrontal cortex. The activity of this area, captured by an fMRI (that's "functional magnetic resonance imaging") machine during learning, predicts whether something that was stored is being recalled in crystal-clear detail. This activity is so reliable that if scientists want to know if you are retrieving something in a robust manner, they don't have to ask you. They can simply look in their machine and see what your left inferior prefrontal cortex is doing.

With this fact in mind, scientist Robert Wagner designed an experiment in which two groups of students were required to memorize a list of words. The first group was shown the words via mass repetition, reminiscent of students cramming for an exam. The second group was shown the words in spaced intervals over a longer period of time, no cramming allowed. In terms of accurate retrieval, the first group fared much worse than the second; activity in the left inferior prefrontal cortex was greatly reduced. These results led Harvard psychology professor Dan Schacter to say: "If you have only one week to study for a final, and only 10 times when you can hit the subject, it is better to space out the 10 repetitions during the week than to squeeze them all together."

Taken together, the relationship between repetition and memory is clear. Deliberately re-expose yourself to the information if you want to retrieve it later. Deliberately re-expose yourself to the information *more elaborately* if you want the retrieval to be of higher quality. Deliberately re-expose yourself to the information more elaborately, and in fixed, spaced intervals, if you want the retrieval to be the most vivid it can be. Learning occurs best when new information is incorporated gradually into the memory store rather than when it is jammed in all at once. So why don't we use such models in our classrooms and boardrooms? Partly it's because educators and business people don't regularly read the *Journal of Neuroscience*. And partly it's because the people who do aren't yet sure which time intervals supply all the magic. Not that timing issues aren't a powerful research focus. In fact, we can divide consolidation into two categories based on timing: fast and slow. To explain how timing issues figure into memory formation, I want to stop for a moment and tell you about how I met my wife.

sparking interest

I was dating somebody else when I first met Kari—and so was she. But I did not forget Kari. She is a physically beautiful, talented,

Emmy-nominated composer, and one of the nicest people I have ever met. When we both became "available" six months later, I immediately asked her out. We had a great time, and I began thinking about her more and more. Turns out she was feeling the same. I asked her out again, and soon we were seeing each other regularly. After two months, it got so that every time we met, my heart would pound, my stomach would flip-flop, and I'd get sweaty palms. Eventually I didn't even have to see her to raise my pulse. Just a picture would do, or a whiff of her perfume, or ... just *music!* Even a fleeting thought was enough to send me into hours of rapture. I knew I was falling in love.

What was happening to effect such change? With increased exposure to this wonderful woman, I became increasingly sensitive to her presence, needing increasingly smaller "input" cues (perfume, for heavens sake?) to elicit increasingly stronger "output" responses. The effect has been long-lasting, with a tenure of almost three decades. Leaving the whys of the heart to poets and psychiatrists, the idea that increasingly limited exposures can result in increasingly stronger responses lies at the heart of how neurons learn things. Only it's not called romance; it's called long-term potentiation.

To describe LTP, we need to leave the high-altitude world of behavioral research and drop down to the more intimate world of cellular and molecular research. Let's suppose you and I are looking at a laboratory Petri dish where two hippocampal neurons happily reside in close synaptic association. I will call the presynaptic neuron the "teacher" and the post-synaptic neuron the "student." The goal of the teacher neuron is to pass on information, electrical in nature, to the student cell. Let's give the teacher neuron some stimulus that inspires the cell to crack off an electrical signal to its student. For a short period of time, the student becomes stimulated and fires excitedly in response. The synaptic interaction between the two is said to be temporarily "strengthened." This phenomenon is termed early LTP.

Unfortunately, the excitement lasts only for an hour or two. If the student neuron does not get the same information from the teacher within about 90 minutes, the student neuron's level of excitement will vanish. The cell will literally reset itself to zero and act as if nothing happened, ready for any other signal that might come its way.

Early LTP is at obvious cross-purposes with the goals of the teacher neuron and, of course, with real teachers everywhere. How does one get that initial excitement to become permanent? Is there a way to transform a student's short-lived response into a long-lived one?

You bet there is: The information must be *repeated* after a period of time has elapsed. If the signal is given only once by the cellular teacher, the excitement will be experienced by the cellular student only transiently. But if the information is repeatedly pulsed in discretely timed intervals (the timing for cells in a dish is about 10 minutes between pulses, done a total of three times), the relationship between the teacher neuron and the student neuron begins to change. Much like my relationship with Kari after a few dates, increasingly smaller and smaller inputs from the teacher are required to elicit increasingly stronger and stronger outputs from the student. This response is termed "late LTP." Even in this tiny, isolated world of two neurons, timed repetition is deeply involved in whether or not learning will occur.

The interval required for synaptic consolidation is measured in minutes and hours, which is why it is called fast consolidation. But don't let this small passage of time disabuse you of its importance. Any manipulation—behavioral, pharmacological, or genetic—that interferes with any part of this developing relationship will block memory formation in total.

Such data provide rock solid evidence that repetition is critical in learning—at least, if you are talking about two neurons in a dish. How about between two people in a classroom? The comparatively simple world of the cell is very different from the complex world of the brain. It is not unusual for a single neuron to have hundreds of synaptic connections with other neurons.

This leads us to a type of consolidation measured in decidedly longer terms, and to stronger end-use implications. It is sometimes called "system consolidation," sometimes "slow consolidation." As we shall see, slow is probably the better term.

a chatty marriage

Nuclear annihilation is a good way to illustrate the differences between synaptic and system consolidation. On August 22, 1968, the Cold War got hot. I was studying history in junior high at the time, living with my Air Force-tethered family at an air base in central Germany, unhappily near ground zero if the atomics were ever to fly in the European theater.

If you could have visited my history class, you wouldn't have liked it. For all the wonderfully intense subject matter—Napoleonic Wars!—the class was taught in a monotonic fashion by a French national, a teacher who really didn't want to be there. And it didn't help my concentration to be preoccupied with the events of the previous day. August 21, 1968, was the morning when a combined contingent of Soviet and Warsaw Pact armies invaded what used to be Czechoslovakia. Our air base went on high alert, and my dad, a member of the U.S. Air Force, had left the evening before. Ominously, he had not yet come home.

The instructor pointed to a large and beautiful painting of the Battle of Austerlitz on the wall, tediously discussing the early wars of Napoleon. I suddenly heard her angry voice say, "Do I need to ask zees twice?" Jolted out of my worried distraction, I turned around to find her looming over my desk. She cleared her throat. "I said, 'Who were Napoleon's enemies in zees battle?' I suddenly realized she had been talking to me, and I blurted out the first words that came to my addled mind. "The Warsaw Pact armies! No? Wait! I mean the Soviet

Union!" Fortunately, the teacher had a sense of humor and some understanding about the day. As the class erupted with laughter, she quickly thawed, tapped my shoulder, and walked back to her desk, shaking her head. "Zee enemies were a coalition of Russian and Austrian armies." She paused. "And Napoleon cleaned their clocks."

Many memory systems are involved in helping me to retrieve this humiliating memory, now almost four decades old. I want to use some of its semantic details to describe the timing properties of system consolidation.

Like Austerlitz, our neurological tale involves several armies of nerves. The first army is the cortex, that wafer-thin layer of nerves that blankets a brain the way an atmosphere blankets a battlefield. The second is a bit of a tongue twister, the medial temporal lobe. It houses another familiar old soldier, the oft-mentioned hippocampus. Crown jewel of the limbic system, the hippocampus helps shape the long-term character of many types of memory. That other teacher-student relationship we were discussing, the one made of neurons, takes place in the hippocampus.

How the cortex and the medial temporal lobe are cabled together tells the story of long-term memory formation. Neurons spring from the cortex and snake their way over to the lobe, allowing the hippocampus to listen in on what the cortex is receiving. Wires also erupt from the lobe and wriggle their way back to the cortex, returning the eavesdropping favor. This loop allows the hippocampus to issue orders to previously stimulated cortical regions while simultaneously gleaning information from them. It also allows us to form memories, and it played a large role in my ability to recount this story to you.

The end result of their association is the creation of long-term memories. How they work to provide stable memories is not well understood, even after three decades of research. We do know something about the characteristics of their communication:

1) Sensory information comes into the hippocampus from

the cortex, and memories form in the cortex by way of the reverse connections.

2) Their electrical marriage starts out being amazingly chatty. Long after the initial stimulus has exited, the hippocampus and the relevant cortical neurons are still yapping about it. Even when I went to bed that night, the hippocampus was busy feeding signals about Austerlitz back to the cortex, replaying the memory over and over again while I slept. This offline processing provides an almost absurdly powerful reason to advocate for regular sleep. The importance of sleep to learning is described in detail in Chapter 7.

3) While these regions are actively engaged, any memory they mediate is labile and subject to amendment. But it doesn't stay that way.

4) After an elapsed period of time, the hippocampus will let go of the cortex, effectively terminating the relationship. This will leave only the cortex holding the memory of the event. But there's an important caveat. The hippocampus will file for cellular divorce only if the cortical memory has first become fully consolidated–only if the memory has changed from transient and amendable to durable and fixed. The process is at the heart of system consolidation, and it involves a complex reorganization of the brain regions supporting a particular memory trace.

So how long does it take for a piece of information, once recruited for long-term storage, to become completely stable? Another way of asking the question is: How long does it take before the hippocampus lets go of its cortical relationship? Hours? Days? Months? The answer surprises nearly everybody who hears it for the first time. The answer is: It can take *years*.

memories on the move

Remember H.M., the fellow whose hippocampus was surgically removed, and along with it the ability to encode new information? H.M. could meet you twice in two hours, with absolutely no recollection of the first meeting. This inability to encode information for long-term storage is called anterograde amnesia. Turns out this famous patient also had retrograde amnesia, a loss of memory of the past. You could ask H.M. about an event that occurred three years before his surgery. No memory. Seven years before his surgery. No memory. If that's all you knew about H.M, you might conclude that his hippocampal loss created a complete memory meltdown. But that's where you'd be wrong. If you asked H.M. about the distant past, say early childhood, he would display a perfectly normal recollection, just as you and I might. He can remember his family, where he lived, details of various events, and so on. This is a conversation with a researcher who studied him for many years:

Researcher: Can you remember any particular event that was special—like a holiday, Christmas, birthday, Easter?

(Now remember, this is a fellow who cannot ever remember meeting this researcher before this interview, though the researcher has worked with him for decades.)

H.M.: There I have an argument with myself about Christmas time.

Researcher: What about Christmas?

H.M.: Well, 'cause my daddy was from the South, and he didn't celebrate down there like they do up here—in the North. Like they don't have the trees or anything like that. And uh, but he came North even though he was born down Louisiana. And I know the name of the town he was born in.

If H.M can recall certain details about his distant past, there must be a point where memory loss began. Where was it? Close analysis revealed that his memory doesn't start to sputter until you get to about the 11th year before his surgery. If you were to graph his memory, you would start out with a very high score and then, 11 years before his surgery, drop it to near zero, where it would remain forever.

What does that mean? If the hippocampus were involved in all memory abilities, its complete removal should destroy all memory abilities—wipe the memory clean. But it doesn't. The hippocampus is relevant to memory formation for more than a decade after the event was recruited for long-term storage. After that, the memory somehow makes it to another region, one not affected by H.M.'s brain losses, and as a result, H.M. can retrieve it. H.M., and patients like him, tell us the hippocampus holds on to a newly formed memory trace for years. Not days. Not months. *Years*. Even a decade or more. System consolidation, that process of transforming a labile memory into a durable one, can take years to complete. During that time, the memory is not stable.

There are, of course, many questions to ask about this process. Where does the memory go during those intervening years? Joseph LeDoux has coined the term "nomadic memory" to illustrate memory's lengthy sojourn through the brain's neural wilderness. But that does not answer the question. Currently nobody knows where it goes, or even *if* it goes. Another question: Why does the hippocampus eventually throw in the towel with its cortical relationships, after spending years nurturing them? Where is the final resting place of the memory once it has fully consolidated? At least in response to that last question, the answer is a bit clearer. The final resting place for the memory is a region that will be very familiar to movie buffs, especially if you like films such as *The Wizard of Oz*, *The Time Machine*, and the original *Planet of the Apes*.

Planet of the Apes was released in 1968, the same year of the Soviet invasion, and appropriately dealt with apocalyptic themes. The main character, a spaceman played by Charleton Heston, had crash-landed onto a planet ruled by apes. Having escaped a gang of malevolent simians at the end of the movie, the last frames show Heston walking along a beach. Suddenly, he sees something off camera of such significance that it makes him drop to his knees. He screams. "You finally did it. God damn you all to hell!" and pounds his fists into the surf, sobbing.

As the camera pulls back from Heston, you see the outline of a vaguely familiar sculpture. Eventually the Statue of Liberty is revealed, half buried in the sand, and then it hits you why Heston is screaming. After this long cinematic journey, he wasn't adventuring on foreign soil. Heston never left Earth. His ending place was the same as his starting place, and the only difference was the timeline. His ship had "crashed" at a point in the far future, a post-apocalyptic Earth now ruled by apes. The first time I encountered data concerning the final resting place of a fully consolidated memory, I immediately thought of this movie.

You recall that the hippocampus is wired to receive information from the cortex as well as return information to it. Declarative memories appear to be terminally stored in the same cortical systems involved in the initial processing of the stimulus. In other words, the final resting place is also the region that served as the initial starting place. The only separation is time, not location. These data have a great deal to say not only about storage but also about recall. Retrieval for a fully mature memory trace 10 years later may simply be an attempt to reconstruct the initial moments of learning, when the memory was only a few milliseconds old! So, the current model looks something like this:

1) Long-term memories occur from accumulations of synaptic changes in the cortex as a result of multiple reinstatements of the memory.

2) These reinstatements are directed by the hippocampus, perhaps for years.

3) Eventually the memory becomes independent of the medial

temporal lobe, and this newer, more stable memory trace is permanently stored in the cortex.

4) Retrieval mechanisms may reconstruct the original pattern of neurons initially recruited during the first moments of learning.

forgetting

Solomon Shereshevskii, a Russian journalist born in 1886, seemed to have a virtually unlimited memory capacity, both for storage and for retrieval. Scientists would give him a list of things to memorize, usually combinations of numbers and letters, and then test his recall. As long as he was allowed 3 or 4 seconds to "visualize" (his words) each item, he could repeat the lists back perfectly, even if the lists had more than 70 elements. He could also repeat the list backward.

In one experiment, a researcher exposed Shereshevskii to a complex formula of letters and numbers containing about 30 items. After a single retrieval test (which Shereshevskii accomplished flawlessly), the researcher put the list in a box, *and waited 15 years.* The scientist then took out the list, found Shereshevskii, and asked him to repeat the formula. Without hesitation, he reproduced the list on the spot, again without error. Shereshevskii's memory of everything he encountered was so clear, so detailed, so *unending*, he lost the ability to organize it into meaningful patterns. Like living in a permanent snowstorm, he saw much of his life as blinding flakes of unrelated sensory information, He couldn't see the "big picture," meaning he couldn't focus on commonalities between related experiences and discover larger, repeating patterns. Poems, carrying their typical heavy load of metaphor and simile, were incomprehensible to him. In fact, he probably could not make sense of the sentence you just read. Shereshevskii couldn't forget, and it affected the way he functioned.

The last step in declarative processing is forgetting. The reason forgetting plays a vital role in our ability to function is deceptively simple. Forgetting allows us to prioritize events. Those events that are irrelevant to our survival will take up wasteful cognitive space if we assign them the same priority as events critical to our survival. So we don't. We insult them by making them less stable. We *forget* them.

There appear to be many types of forgetting, categories cleverly enumerated by Dan Schacter, the father of research on the phenomenon, in his book *The Seven Sins of Memory*. Tip-of-the-tongue issues, absent-mindedness, blocking habits, misattribution, biases, suggestibility—the list reads like a cognitive Chamber of Horrors for students and business professionals alike. Regardless of the type of forgetting, they all have one thing in common. They allow us to drop pieces of information in favor of others. In so doing, forgetting helped us to conquer the Earth.

ideas

How can we use all of this information to conquer the classroom? The boardroom? Exploring the timing of information re-exposure is one obvious arena where researchers and practitioners might do productive work together. For example, we have no idea what this means for marketing. How often must you repeat the message before people buy a product? What determines whether they still remember it six months later, or a year later?

Minutes and hours

The day of a typical high-school student is segmented into five or six 50-minute periods, consisting of unrepeated (and unrelenting) streams of information. Using as a framework the timing requirements suggested by working memory, how would you change this five-period fire hose? What you'd come up with might be the strangest classroom experience in the world. Here's my fantasy:

In the school of the future, lessons are divided into 25-minute modules, cyclically repeated throughout the day. Subject A is taught for 25 minutes, constituting the first exposure. Ninety minutes later, the 25-minute content of Subject A is repeated, and then a third time. *All* classes are segmented and interleaved in such a fashion. Because these repetition schedules slow down the amount of information capable of being addressed per unit of time, the school year is extended into the summer.

Days and weeks

We know from Robert Wagner that multiple reinstatements provide demonstrable benefit over periods of days and even weeks.

In the future school, every third or fourth day would be reserved for reviewing the facts delivered in the previous 72 to 96 hours. During these "review holidays," previous information would be presented in compressed fashion. Students would have a chance to inspect the notes they took during the initial exposures, comparing them with what the teacher was saying in the review. This would result in a greater elaboration of the information, and it would help the teachers deliver accurate information. A formalized exercise in error-checking soon would become a regular and positive part of both the teacher and student learning experiences.

It is quite possible that such models would eradicate the need for homework. At its best, homework served only to force the student to repeat content. If that repetition were supplied during the course of the day, there might be little need for further re-exposure. This isn't because homework isn't important as a concept. In the future school, it may simply be unnecessary.

Could models like these actually work? Deliberately spaced repetitions have not been tested rigorously in the real world, so there are lots of questions. Do you really need three separate repetitions per subject per day to accrue a positive outcome? Do all subjects need such repetition? Might such interleaved vigor *hurt* learning, with constant repetitions beginning to interfere with one another as the day wore on? Do you really need review holidays, and if so, do you need them every three to four days? We don't know.

Years and years

Today, students are expected to know certain things by certain grades. Curiously absent from this model is how durable that learning remains after the student completes the grade. Given that system consolidation can take *years*, might the idea of grade-level expectations need amending? Perhaps learning in the long view should be thought of the same way one thinks of immune booster shots, with critical pieces of information being repeated on a yearly or semi-yearly basis.

In my fantasy class, this is exactly what happens. Repetitions begin with a consistent and rigorous review of multiplication tables, fractions, and decimals. First learned in the third grade, six-month and yearly review sessions on these basic facts occur through sixth grade. As mathematical competencies increase in sophistication, the review content is changed to reflect greater understanding. *But the cycles are still in place.* In my fantasy, these consistent repetition disciplines, stretched out over long periods of time, create enormous benefits for every academic subject, especially foreign languages.

You've probably heard that many corporations, especially in technical fields, are disappointed by the quality of the American undergraduates they hire. They have to spend money retraining many of their newest employees in certain basic skills that they often think should have been covered in college. One of my business fantasies would partner engineering firms with colleges of engineering. It involves shoring up this deficit by instituting post-graduate repetition experiences. These reinstatement exercises would be instituted the week after graduation and continue through the first year of employment. The goal? To *review every important technical subject relevant to the employee's new job.* Research would establish not only the choice of topics to be reviewed but also the optimal spacing of the repetition.

My fantasy shares the teaching load between firm members and the academic community, extending the idea of a bachelor's degree into the workplace. This hybridization aligns business professionals with researchers, ensuring that companies have exposure to the latest advances in their fields (and informing researchers on the latest practical day-to-day issues business professionals face). In my fantasy, the program becomes so popular that the more experienced engineers also begin attending these refresher courses, inadvertently rubbing shoulders with younger generations. The old guard is surprised by how much they have forgotten, and how much the review and cross-hybridization, both with research professionals and younger students, aid their own job performance.

I wish I could tell you this all would work, but instead all I can say is that memory is not fixed at the moment of learning, and repetition provides the fixative.

Summary

Rule #6
Remember to repeat.

* Most memories disappear within minutes, but those that survive the fragile period strengthen with time.

* Long-term memories are formed in a two-way conversation between the hippocampus and the cortex, until the hippocampus breaks the connection and the memory is fixed in the cortex—which can take years.

* Our brains give us only an approximate view of reality, because they mix new knowledge with past memories and store them together as one.

* The way to make long-term memory more reliable is to incorporate new information gradually and repeat it in timed intervals.

Get more at www.brainrules.net

sleep

Sleep well, think well.

IT'S NOT THE MOST comfortable way to earn an entry in the Guinness Book of World Records, obtain an A on a high-school science-fair project, and meet a world-famous scientist. In 1965, 17-year-old Randy Gardner decided that his science-fair project would involve not sleeping for 11 straight days and observing what happened. To the astonishment of just about everyone, he accomplished the feat, setting a world record that year for sleep loss. The project attracted the attention of scientist William Dement, who was given permission to study what happened to the teenager's mind during the week and a half he was awake.

What happened to Randy's mind was extraordinary. To put it charitably, it started to malfunction. In short order, he became irritable, forgetful, nauseous, and, to no one's surprise, unbelievably tired. Five days into his experiment, Randy began to suffer from what could pass for Alzheimer's disease. He was actively hallucinating, severely disoriented, and paranoid. He thought a local radio host was out to get him because of his changes in memory. In the last four days of his experiment, he lost motor function, his fingers trembling and his speech slurred. Curiously, on the final day, he still was able to beat Dement at pinball, doing so 100 consecutive times.

Some unfortunate souls don't have the luxury of experimenting. They become suddenly—and permanently—incapable of ever going to sleep again. Fatal Familial Insomnia is one of the rarest human genetic disorders that exists, affecting only about 20 families worldwide. That rarity is a blessing, because the disease follows a course straight through mental-health hell. In middle to late adulthood, the person begins to experience fevers, tremors, and profuse sweating. As the insomnia becomes permanent, these symptoms are accompanied by increasingly uncontrollable muscular jerks and tics. The person soon experiences crushing feelings of depression and anxiety. He or she becomes psychotic. Finally, mercifully, the patient slips into a coma and dies.

So we know bad things happen when we don't get any sleep. But, considering that sleep occupies a walloping one-third of our time on the planet, it is incredible to contemplate that we still don't know *why* we need to sleep. Not that there haven't been clues. One strong hint came about 10 years ago, from a group of researchers who left a bunch of wires stuck inside a rat's brain. The rat had just learned to negotiate a maze when it decided to take a nap. The recording device was still attached to those wires, and it was still on. But to understand how this relates to the purpose of sleep, let's look at what the brain is doing while we sleep.

you call this rest?

If you ever get a chance to listen in on a living brain while it is slumbering, you'll have to get over your disbelief. The brain does not appear to be asleep at all. Rather, it is almost unbelievably active during "rest," with legions of neurons crackling electrical commands to one another in constantly shifting patterns—displaying greater rhythmical activity during sleep, actually, than when it is wide awake.

The only time you can observe a real resting period for the brain (where the amount of energy consumed is less than during a similar awake period) is in the deepest parts of what is called non-REM sleep. But that takes up only about 20 percent of the total sleep cycle, which is why researchers early on began to disabuse themselves of the notion that the reason we rest is so that we can rest. When the brain is asleep, the brain is not resting at all.

Even so, most people report that sleep is powerfully restorative, and they point to the fact that if they don't get enough sleep, they don't think as well. That is measurably true, as we shall see shortly. And so we find ourselves in a quandary: Given the amount of energy the brain is using, it seems impossible that you could receive anything approaching mental rest and restoration during sleep.

Even if the brain doesn't behave itself bioenergetically, other parts of the body do rest during sleep, in something like a human version of micro-hibernation. That introduces a second puzzle: Sleep makes us exquisitely vulnerable to predators. Indeed, deliberately going off to dreamland unprotected in the middle of a bunch of hostile hunters (such as leopards, our evolutionary roommates in eastern Africa) seems like a behavior dreamed up by our worst enemies. There must be something terribly important we need to accomplish during sleep if we are willing to take such risks in order to get it. Exactly what is it that is so darned important?

The scientist who studied sleepless Randy Gardner made a substantial early contribution to answering such questions. Often called the father of sleep research, Dement is a white-haired man with a broad smile who at this writing is in his late 70s. He says pithy things about our slumbering habits, such as "Dreaming permits each and every one of us to be quietly and safely insane every night of our lives."

Dement studied many aspects of the human sleep cycle. What he began to uncover was this: "Sleeping" brains, like soldiers on a battlefield, are actually locked in vicious, biological combat. The conflict involves a pitched battle between two powerful and opposing drives, each made of legions of brain cells and biochemicals with very different agendas. Though localized in the head, the theater of operations for these armies engulfs every corner of the body. This fight is sometimes referred to as the "opponent process" model.

As Dement began to define these two opposing drives, he noticed some strange things about the war they were waging. First, these forces are not engaged just during the night, while we sleep, but also during the day, while we are awake. Second, they are doomed to a combat schedule in which each army sequentially wins one battle, then promptly loses the next battle, then quickly wins the next and so on, cycling through this win/loss column every day and every night. The third strange thing is that no one army ever claims final victory in this war. This incessant engagement results in the cyclical waking and sleeping experiences all humans encounter every day (and night) of our lives.

Dement was not working in isolation. His mentor, a gifted researcher named Nathaniel Kleitman, gave him many of his initial insights. If Dement can be considered the father of sleep research, Kleitman certainly could qualify as its grandfather. An intense Russian man with bushy eyebrows, Nathaniel Kleitman may be best noted for his willingness to experiment not only on himself but also on his children. When it appeared that a colleague of his had discovered Rapid Eye Movement (REM) sleep, Kleitman promptly volunteered his daughter for experimentation, and she just as promptly confirmed the finding. But one of the most interesting experiments of Kleitman's long career occurred in 1938, when he persuaded a colleague to join him 150 feet underground in Mammoth Cave in Kentucky for an entire month.

Free of sunlight and daily schedules, Kleitman could ask whether the routines of wakefulness and sleep cycled themselves automatically through the human body. His observations were mixed, but the experiment provided the first real hint that such an automatic device did exist in our bodies. Indeed, we now know that the body possesses a series of internal clocks, all controlled by discrete regions in the brain, providing a regular rhythmic schedule to our waking and sleeping experiences. This is surprisingly similar to the buzzing of a wristwatch's internal quartz crystal. An area of the brain called the suprachiasmatic nucleus, part of that hypothalamus we discussed earlier, appears to contain just such a timing device. Of course, we have not been characterizing these pulsing rhythms as a benign wristwatch. We have been characterizing them as a violent war. One of Kleitman's and Dement's greatest contributions was to show that this nearly automatic rhythm occurs as a result of the continuous conflict between two opposing forces.

With the idea that such forces are under internal control, we can explore them in greater detail, beginning with a description of their names. One army is composed of neurons, hormones, and various other chemicals that do everything in their power to keep you awake. This army is called the circadian arousal system (often referred to simply as "process C"). If this army had its way, it would make you stay up all the time. Fortunately, it is opposed by an equally powerful army, also made of brain cells, hormones, and various chemicals. These combatants do everything in their power to put you to sleep. They are termed the homeostatic sleep drive ("process S"). If this army had its way, you would go to sleep and never wake up.

It is a strange, even paradoxical, war. The longer one army controls the field, for example, the more likely it is to lose the battle. It's almost as if each army becomes exhausted from having its way and eventually waves a temporary white flag. Indeed, the longer you are awake (the victorious process C doing victory laps around your head), the greater the probability becomes that the circadian arousal system will eventually cede the field to its opponent. You then go to sleep. For most people, this act of capitulation comes after about 16 hours of active consciousness. This will occur even if you are living in a cave.

Conversely, the longer you are asleep (the triumphant process S now doing the heady victory laps), the greater the probability becomes that the homeostatic sleep drive will similarly cede the field to *its* opponent, which is, of course, the drive to keep you awake. The result of this surrender is that you wake up. For most people, the length of time prior to capitulation is about half of its opponent's, about eight hours of blissful sleep. And this also will occur even if you are living in a cave.

Except for the unfortunate members of 20 or so families worldwide, Kleitman, Dement, and a host of other researchers were able to show that such dynamic tension is a normal—even critical—part of our daily lives. In fact, the circadian arousal system and the homeostatic sleep drive are locked in a daily warfare of victory and surrender so predictable, you can actually graph it. Stated formally, process S maintains the duration and intensity of sleep, while process C determines the tendency and timing of the need to go to sleep.

Now, this war between the two armies does not go unsupervised. Internal and external forces help regulate the conflict, defining for us both the amount of sleep we need and the amount of sleep we get. We will focus on two of the internal forces, chronotype and the nap zone. To understand how these work, we must leave the intricacies of battle for a moment and explore instead the life of newspaper cartoonists and advice columnists. Oh, and we will also talk about birds.

lark or owl?

The late advice columnist Ann Landers would vehemently declare, "No one's going to call me until I'm ready!" and then take her phone off the hook between 1 and 10 a.m. Why? This was the time she normally went to sleep. The cartoonist Scott Adams, creator of the comic strip *Dilbert*, never would think of starting his day at 10 a.m. "I'm quite tuned into my rhythms," he has said. "I never try to do any creating past noon. ... I do the strip from 6 to 7 a.m." Here we have two creative and well-accomplished professionals, one who starts working just as the other's workday is finished.

About 1 in 10 of us is like *Dilbert's* Adams. The scientific literature calls such people larks (more palatable than the proper term, "early chronotype"). In general, larks report being most alert around noon and feel most productive at work a few hours before they eat lunch. They don't need an alarm clock, because they invariably get up before the alarm rings—often before 6 a.m. Larks cheerfully report their favorite mealtime as breakfast and generally consume much less coffee than non-larks. Getting increasingly drowsy in the early evening, most larks go to bed (or want to go to bed) around 9 p.m.

Larks are the mortal enemy of the 2 in 10 humans who lie at the other extreme of the sleep spectrum: "late chronotypes," or owls. In general, owls report being most alert around 6 p.m., experiencing their most productive work times in the late evening. They rarely want to go to bed before 3 a.m. Owls invariably need an alarm clock to get them up in the morning, with extreme owls requiring multiple alarms to ensure arousal. Indeed, if owls had their druthers, most would not wake up much before 10 a.m. Not surprisingly, late chronotypes report their favorite mealtime as dinner, and they would drink gallons of coffee all day long to prop themselves up at work if given the opportunity. If it sounds to you as though owls do not sleep as well as larks in our society, you are right on the money. Indeed, late chronotypes usually accumulate a massive "sleep debt" as they go through life.

The behaviors of larks and owls are very specific. Researchers think these patterns are detectable in early childhood and burned into the genetic complexities of the brain that govern our sleep/wake cycle. At least one study shows that if Mom or Dad is a lark, half of their kids will be, too. Larks and owls cover only about 30 percent of the population. The rest of us are called hummingbirds. True to the idea of a continuum, some hummingbirds are more owlish, some are more larkish, and some are in between. To my knowledge, no birdish moniker has ever been applied to those people who seem to need only four or five hours of sleep. They instead are referred to as suffering from "healthy insomnia."

So how much sleep does a person need? Given all of our recent understanding about how and when we sleep, you might expect that scientists would come up with the answer fairly quickly. Indeed, they have. The answer is: We don't know. You did not read that wrong. After all of these centuries of experience with sleep, we still don't know how much of the stuff people actually need. Generalizations don't work: When you dig into the data on humans, what you find is not remarkable uniformity but remarkable individuality. To make matters worse, sleep schedules are unbelievably dynamic. They change with age. They change with gender. They change depending upon whether or not you are pregnant, and whether or not you are going through puberty. There are so many variables one must take into account that it almost feels as though you've asked the wrong question. So let's invert the query. How much sleep *don't* you need? In other words, what are the numbers that disrupt normal function? That turns out to be an important question, because it is possible to become dysfunctional with too much sleep *or* not enough. Whatever amount of sleep is right for you, when robbed of that (in either direction), bad things really do happen to your brain.

napping in the free world

Given that sleep rhythms fight their battles 24 hours a day, researchers have studied the skirmishes occurring not only in the night but also in the day. One area of interest is the persistent need to take a nap, and to do so at very specific times of the day.

It must have taken some getting used to, if you were a staffer in the socially conservative early 1960s. Lyndon Baines Johnson, 36th president of the United States and leader of the free world, routinely closed the door to his office in the midafternoon and put on his pajamas. He then proceeded to take a 30-minute nap. Rising refreshed, he would tell aides that such a nap gave him the stamina to work the long hours required of the U.S. commander-in-chief during the Cold War. Such presidential behavior might seem downright weird. But if you ask sleep researchers like William Dement, his response might surprise you: It was LBJ who was acting normally; the rest of us, who refuse to bring our pajamas to work, are the abnormal ones. And Dement has a fair amount of data to back him up.

LBJ was responding to something experienced by nearly everyone on the planet. It goes by many names—the midday yawn, the post-lunch dip, the afternoon "sleepies." We'll call it the nap zone, a period of time in the midafternoon when we experience transient sleepiness. It can be nearly impossible to get anything done during this time, and if you attempt to push through, which is what most of us do, you can spend much of your afternoon fighting a gnawing tiredness. It's a fight because the brain really wants to take a nap and doesn't care what its owner is doing. The concept of "siesta," institutionalized in many other cultures, may have come as an explicit reaction to the nap zone.

At first, scientists didn't believe the nap zone existed except as an artifact of sleep deprivation. That has changed. We now know that some people feel it more intensely than others. We know it is not related to a big lunch (although a big lunch, especially one loaded with carbs, can greatly increase its intensity). It appears, rather, to be a part of our evolutionary history. Some scientists think that a long sleep at night and a short nap during the midday represent human sleep behavior at its most natural.

When you chart the process S curve and process C curve, you can see that they flat-line in the same place—in the afternoon. Remember that these curves are plotting the progress of a war between two opposed groups of cells and biochemicals. The battle clearly has reached a climactic stalemate. An equal tension now exists between the two drives, which extracts a great deal of energy to maintain. Some researchers, though not all, think this equanimity in tension drives the nap zone. Regardless, the nap zone matters, because our brains don't work as well during it. If you are a public speaker, you already know it is darn near fatal to give a talk in the midafternoon. The nap zone also is literally fatal: More traffic accidents occur during it than at any other time of the day.

On the flip side, one NASA study showed that a 26-minute nap improved a pilot's performance by more than 34 percent. Another study showed that a 45-minute nap produced a similar boost in cognitive performance, lasting more than six hours. Still other researchers demonstrated that a 30-minute nap taken prior to staying up all night can prevent a significant loss of performance during that night.

If that's what a nap can do, imagine the benefits of a full night's sleep. Let's look at what can happen when we ignore these internal forces, and when we embrace them.

go ahead, sleep on it

If central casting ever called you to suggest a character in history representing the archetypal brilliant-but-mad-looking scientist, Dimitri Ivanovich Mendeleyev might be in your top five list. Hairy and opinionated, Mendeleyev possessed the lurking countenance of a Rasputin, the haunting eyes of Peter the Great, and the moral flexibility of both. He once threatened to commit suicide if a young lady didn't marry him. She consented, which was quite illegal, because, unbeknownst to the poor girl, Mendeleyev was already married. This trespass kept him out of the Russian Academy of Sciences for a while, which in hindsight may have been a bit rash, as Mendeleyev single-handedly systematized the entire science of chemistry.

His Periodic Table of the Elements—a way of organizing every atom that had so far been discovered—was so prescient, it allowed room for all the elements yet to be found and even predicted some of their properties. But what's most extraordinary is this: Mendeleyev says he first came up with the idea in his sleep. Contemplating the nature of the universe while playing solitaire one evening, he nodded off. When he awoke, he knew how all of the atoms in the universe were organized, and he promptly created his famous table. Interestingly, he organized the atoms in repeating groups of seven.

Mendeleyev is hardly the only scientist who has reported feelings of inspiration after having slept, of course. Is there something to the notion of "Let's sleep on it"? What's the relationship between ordinary sleep and extraordinary learning?

Mountains of data demonstrate that a healthy sleep can indeed boost learning significantly, in certain types of tasks. These results generate a great deal of interest among sleep scientists and, unsurprisingly, no small amount of controversy. How should we define learning, they debate; exactly what is improvement? But there are many examples of the phenomenon. One study stands out in particular.

Students were given a series of math problems and prepped with a method to solve them. The students weren't told there was also an easier, "shortcut" way to solve the problems, potentially discoverable while doing the exercise. The question was: Is there any way to jumpstart, even speed up, their insights? Can you get them to put this other method on their radar screens? The answer was yes, *if you allow them to sleep on it.* If you let 12 hours pass after the initial training and ask the students to do more problems, about 20 percent will have discovered the shortcut. But, if in that 12 hours you also allow eight or so hours of regular sleep, that figure triples to about 60 percent. No matter how many times the experiment is run, the sleep group consistently outperforms the non-sleep group about 3 to 1.

Sleep has been shown to enhance tasks that involve visual texture discrimination, motor adaptations, and motor sequencing. The type of learning that appears to be most sensitive to sleep improvement is that which involves learning a procedure. Simply disrupt the night's sleep at specific stages and retest in the morning, and you eliminate any overnight learning improvement. Clearly, for specific types of intellectual skill, sleep can be a great friend to learning.

sleep loss = brain drain

It won't surprise you, then, that lack of sleep hurts learning. In fact, a highly successful student can be set up for a precipitous academic fall, just by adjusting the number of hours she sleeps. Take an A student used to scoring in the top 10 percent of virtually anything she does. One study showed that if she gets just under seven hours of sleep on weekdays, and about 40 minutes more on weekends, she will begin to score in the bottom 9 percent of non-sleep-deprived individuals. Cumulative losses during the week add up to cumulative deficits during the weekend—and, if not paid for, that sleep debt will be carried into the next week.

Another study followed soldiers responsible for operating complex military hardware. One night's loss of sleep resulted in about a 30 percent loss in overall cognitive skill, with a subsequent drop in performance. Bump that to two nights' loss, and the figure becomes 60 percent. Other studies extended these findings. When sleep was restricted to six hours or less per night for just five nights, for example, cognitive performance matched that of a person suffering from 48 hours of continual sleep deprivation.

More recent research has begun to shed light on other functions that do not at first blush seem associated with sleep. When people become sleep-deprived, for example, their ability to utilize the food they are consuming falls by about one-third. The ability to make insulin and to extract energy from the brain's favorite dessert, glucose, begins to fail miserably. At the same time, you find a marked need to have more of it, because the body's stress hormone levels begin to rise in an increasingly deregulated fashion. If you keep up the behavior, you appear to accelerate parts of the aging process. For example, if healthy 30-year-olds are sleep-deprived for six days (averaging, in this study, about four hours of sleep per night), parts of their body chemistry soon revert to that of a 60-year-old. And if they are allowed to recover, it will take them almost a week to get back to their 30-year-old systems.

The bottom line is that sleep loss means mind loss. Sleep loss cripples thinking, in just about every way you can measure thinking. Sleep loss hurts attention, executive function, immediate memory, working memory, mood, quantitative skills, logical reasoning ability, general math knowledge. Eventually, sleep loss affects manual dexterity, including fine motor control (except, perhaps, for pinball) and even gross motor movements, such as the ability to walk on a treadmill.

When you look at all of the data combined, a consistency emerges: Sleep is rather intimately involved in learning. It is observable with large amounts of sleep; it is observable with small amounts of sleep; it is observable all the time. Of course, explaining exactly *how* sleep improves performance has not been as easy as demonstrating the fact *that* it improves performance. Given the importance of the issue to the Brain Rule, let's try anyway.

Consider the following true story of a successfully married, incredibly detail-oriented accountant. Even though dead asleep, he regularly gives financial reports to his wife all night long. Many of these reports come from the day's activities. (Incidentally, if his wife wakes him up—which is often, because his financial broadcasts are loud—the accountant becomes amorous and wants to have sex.) Are we all organizing our previous experiences while we sleep? Could this not only explain all of the other data we have been discussing, but also finally give us a reason why we sleep?

To answer these questions, we must return to our story of the hapless rat who, 10 years ago, was unfortunate to have fallen asleep with a bunch of wires stuck inside his brain. The "wires" are electrodes placed near individual neurons. Hook these electrodes up to a recording device, and you can eavesdrop on the brain while it is talking to itself, something like a CIA phone tap, listening to the individual chatter of neurons as they process information. Even in a tiny rat's brain, it is not unusual these days to listen in on up to 500 neurons at once. So what are they all saying? If you listen in while the rat is acquiring new information, like learning to navigate a maze, you soon will detect something extraordinary. A very discrete "maze-specific" pattern of electrical stimulation begins to emerge. Working something like the old Morse code, a series of neurons begin to crackle in a specifically timed sequence during the learning. Afterward, the rat will always fire off that pattern whenever it travels through the maze. It appears to be an electrical representation of the rat's new maze-navigating thought patterns (at least, as many as 500 electrodes can detect).

When the rat goes to sleep, it begins to *replay the maze-pattern sequence*. The animal's brain replays what it learned while it slumbers, reminiscent of our accountant. Always executing the pattern in a specific stage of sleep, the rat repeats it over and over again—and much faster than during the day. The rate is so furious, the sequence is replayed thousands of times. If a nasty graduate student decides to wake up the rat during this stage, called slow-wave sleep, something equally extraordinary is observed. The rat has trouble remembering the maze the next day. Quite literally, the rat seems to be consolidating the day's learning the night *after* that learning occurred, and an interruption of that sleep disrupts the learning cycle.

This naturally caused researchers to ask whether the same was true for humans. The answer? Not only do we do such processing, but we do it in a far more complex fashion. Like the rat, humans appear to replay certain daily learning experiences at night, during the slow-wave phase. But unlike the rat, more emotionally charged memories appear to replay at a different stage in the sleep cycle.

These findings represent a bombshell of an idea: Some kind of offline processing is occurring at night. Is it possible that the reason we need to sleep is simply to shut off the exterior world for a while, allowing us to divert more attentional resources to our cognitive interiors? Is it possible that the reason we need to sleep is so that we can learn?

It sounds compelling, but of course the real world of research is much messier. Many findings appear to complicate, if not fully contradict, the idea of offline processing. For example, brain-damaged individuals who lack the ability to sleep in the slow-wave phase nonetheless have normal, even improved, memory. So do individuals whose REM sleep is suppressed by antidepressant medications. Exactly how to reconcile these data with the previous findings is a subject of intense scientific debate. What's always needed is more research—but not just at the lab bench.

ideas

What if businesses and schools took the sleep needs of their employees and students seriously? What would a modern office building look like? What would a school look like? These are not idle questions. The effects of sleep deprivation are thought to cost U.S. businesses more than $100 billion a year. I have a few ideas ripe for real-world research.

Match chronotypes

A number of behavioral tests can discriminate larks from owls from hummingbirds fairly easily. And given advances in genetic research, you may in the future need only a blood test to characterize your process C/process S graphs. The bottom line is, we can determine the hours when a person is likely to experience his or her major productivity peaks.

Here's an obvious idea: What if we began to match chronotypes to work schedules? Twenty percent of the workforce is already at sub-optimal productivity in the current 9-to-5 model. What if we created several work schedules, based on the chronotypes of the employees? We might gain more productivity and a greater quality of life for those unfortunate employees who otherwise are doomed to carry a permanent sleep debt. We might get more productive use out of our buildings if they remained open instead of lying dormant half the night. A business of the future will need to become involved in some aspect of its employees' sleep schedules.

We could do the same in education. Teachers are just as likely to be late chronotypes as their students. Why not put them together? Would you increase the competencies of the teacher? The students? Free of the nagging consequences of their sleep debts, their educational experiences might become more robust simply because each was more fully capable of mobilizing his God-given IQ.

Variable schedules also would take advantage of the fact that sleep needs change throughout a person's life span. For example, data suggest that students temporarily shift to more of an owlish chronotype as they transit through their teenage years. This has led some school districts to start their high-school classes after 9 a.m. This may make some sense. Sleep hormones (such as the protein melatonin) are at their maximum levels in the teenage brain. The natural tendency of these kids is to sleep more, especially in the morning. As we age, we tend to get less sleep, and some evidence suggests we need less sleep, too. An employee who starts out with her greatest productivity in one schedule may, as the years go by, keep a similar high level of output simply by switching to a new schedule.

Promote naps

To embrace the midday nap zone, engineers at MetroNaps have created a nap-on-the-go device called a Sleep Pod. "It looks like a sperm that got electrocuted!" exclaimed one person upon seeing the device for the first time. Actually, the pods are portable glorified recliners that can fit in an office—complete with light-canceling visors, noise-canceling earphones, heat-canceling circulation coils, and—at more than $14,000 each—budget-canceling prices. The company, based in New York, has pods in four countries and is busy expanding its business. Others are bringing naps into the workplace, too. Hotels with stacked-bed "nap salons" have sprung up all over Japan. A Boston-based researcher named William Anthony is trying to create National Napping Day, a day set aside so that everybody can take a nap. He finds that 70 percent of Americans who admit to being workplace nappers still have to take their naps in secret. The favored clandestine venue? In the back seat of the employee's car. At lunch.

What if businesses and schools took seriously the existence of nap zones? No meetings or classes would ever be scheduled at the time when the process C and process S curves are flat-lined. No high-demand presentations and no critical exams would be assigned anywhere near the collision of these two curves. Instead, there would be deliberately planned downshifts. Naps would be accorded the same deference that businesses reluctantly treat lunch, or even potty breaks: a necessary nod to an employee's biological needs. Companies would create a designated space for employees to take one half-hour nap each workday. The advantage would be straightforward. People hired for their intellectual strength would be allowed to keep that strength in tip-top shape. "What other management strategy will improve people's performance 34 percent in just 26 minutes?" says Mark Rosekind, the NASA scientist who conducted that eye-opening research on naps and pilot performance.

Try sleeping on it

Given the data about a good night's rest, organizations might tackle their most intractable problems by having the entire "solving team" go on a mini-retreat. Once arrived, employees would be presented with the problem and asked to think about solutions. But they would not start coming to conclusions, or even begin sharing ideas with each other, before they had slept about eight hours. When they awoke, would the same increase in problem-solving rates available in the lab also be available to that team? We ought to find out.

Summary

Rule #7
Sleep well, think well.

* The brain is in a constant state of tension between cells and chemicals that try to put you to sleep and cells and chemicals that try to keep you awake.

* The neurons of your brain show vigorous rhythmical activity when you're asleep—perhaps replaying what you learned that day.

* People vary in how much sleep they need and when they prefer to get it, but the biological drive for an afternoon nap is universal.

* Loss of sleep hurts attention, executive function, working memory, mood, quantitative skills, logical reasoning, and even motor dexterity.

Get more at www.brainrules.net

stress

Rule #8
Stressed brains don't learn the same way.

IT IS, BY ANY measure, a thoroughly rotten experiment.

Here is this beautiful German shepherd, lying in one corner of a metal box, whimpering. He is receiving painful electric shocks, stimuli that should leave him howling in pain. Oddly enough, the dog could easily get out. The other side of the box is perfectly insulated from shocks, and only a low barrier separates the two sides. Though the dog could jump over to safety when the whim strikes him, the whim doesn't strike him. Ever. He just lies down in the corner of the electric side, whimpering with each jarring jolt. He must be physically removed by the experimenter to be relieved of the experience.

What has happened to that dog?

A few days before entering the box, the animal was strapped to a restraining harness rigged with electric wires, inescapably receiving the same painful shock day and night. And at first he didn't just stand there taking it, he *reacted*. He howled in pain. He urinated. He strained mightily against his harness in an increasingly desperate attempt to link some behavior of his with the cessation of the pain. But it was no use. As the hours and even days ticked by, his resistance eventually subsided. Why? The dog began to receive a very clear message: The pain was not going to stop; the shocks were going to be forever. *There was no way out.* Even after the dog had been released from the harness and placed into the metal box with the escape route, he could no longer understand his options. Indeed, most learning had been shut down, and that's probably the worst part of all.

Those of you familiar with psychology already know I am describing a famous set of experiments begun in the late 1960s by legendary psychologist Martin Seligman. He coined the term "learned helplessness" to describe both the perception of inescapability and its associated cognitive collapse. Many animals behave in a similar fashion where punishment is unavoidable, and that includes humans. Inmates in concentration camps routinely experienced these symptoms in response to the horrid conditions of the internment, and some camps even gave it the name *Gamel*, derived from the colloquial German word *Gameln*, which literally means "rotting." Perhaps not surprisingly, Seligman has spent the balance of his career studying how humans respond to optimism.

What is so awful about severe, chronic stress that it can wreak such extraordinary changes in behavior? Why is learning so radically altered? Let's begin with a definition of stress, talk about biological responses, and then move to the relationship between stress and learning. Along the way, we will talk about marriage and parenting, about the workplace, and about the first and only time I ever heard my mother, a fourth-grade teacher, swear. It was her first real encounter with learned helplessness.

terror and titillation

We begin with an attempt at definitions, and, as is true of all things cognitive, we suddenly run into turbulence. First, not all stress is the same. Certain types of stress really hurt learning, but some types of stress *boost* learning. Second, it's difficult to detect when someone is experiencing stress. Some people love skydiving for recreation; it's others' worst nightmare. Is jumping out of an airplane inherently stressful? The answer is no, and that highlights the subjective nature of stress.

The body isn't of much help in providing a definition, either. There is no unique grouping of physiological responses capable of telling a scientist whether or not you are experiencing stress. The reason? Many of the same mechanisms that cause you to shrink in horror from a predator are also used when you are having sex—or even while you are consuming your Thanksgiving dinner. To your body, saber-toothed tigers and orgasms and turkey gravy look remarkably similar. An aroused physiological state is characteristic of both stress and pleasure.

So what's a scientist to do? A few years ago, gifted researchers Jeansok Kim and David Diamond came up with a three-part definition that covers many of the bases. In their view, if all three are happening simultaneously, a person is stressed.

Part one: There must be an aroused physiological response to the stress, and it must be measurable by an outside party. I saw this in obvious fashion the first time my then 18-month-old son encountered a carrot on his plate at dinner. He promptly went ballistic: He screamed and cried and peed in his diaper. His aroused physiological state was immediately measurable by his dad, and probably by anyone else within a half mile of our kitchen table.

Part two: The stressor must be perceived as aversive. This can be assessed by a simple question: "If you had the ability to turn down the severity of this experience, or avoid it altogether, would you?" It was obvious where my son stood on the matter. Within seconds, he took the carrot off his plate and threw it on the floor. Then he deftly got down off his chair and tried to stomp on the predatory vegetable. The avoidance question was answered in full.

Part three: The person must not feel in control of the stressor.

Like a volume knob on some emotional radio, the more the loss of control, the more severe the stress is perceived to be. This element of control and its closely related twin, predictability, lie at the heart of learned helplessness. My son reacted as strongly as he did in part because he knew I wanted him to eat the carrot, and he was used to doing what I told him to do. Control was the issue. Despite my picking up the carrot, washing it, then rubbing my tummy while enthusiastically saying "yum, yum," he was having none of it. Or, more important, he was wanting to have none of it, and he thought I was going to make him have all of it. Out-of-control carrot equaled out-of-control behavior.

When you find this trinity of components working together, you have the type of stress easily measurable in a laboratory setting. When I talk about stress, I am usually referring to situations like these.

flooding the system

You can feel your body responding to stress: Your pulse races, your blood pressure rises, and you feel a massive release of energy. That's the famous hormone adrenaline at work. It's spurred into action by your brain's hypothalamus, that pea-size organ sitting almost in the middle of your head. When your sensory systems detect stress, the hypothalamus reacts by sending a signal to your adrenal glands, lying far away on the roof of your kidneys. The glands immediately dump bucketloads of adrenaline into your bloodstream. The overall effect is called the fight or flight response.

But there's a less famous hormone at work, too—also released by the adrenals, and just as powerful as adrenalin. It's called cortisol. You can think of it as the "elite strike force" of the human stress response. It's the second wave of our defensive reaction to stressors, and, in small doses, it wipes out most unpleasant aspects of stress, returning us to normalcy.

Why do our bodies need to go through all this trouble? The

answer is very simple. Without a flexible, immediately available, highly regulated stress response, we would die. Remember, the brain is the world's most sophisticated survival organ. All of its many complexities are built toward a mildly erotic, singularly selfish goal: to live long enough to thrust our genes on to the next generation. Our reactions to stress serve the live-long-enough part of that goal. Stress helps us manage the threats that could keep us from procreating.

And what kinds of sex-inhibiting threats did we experience in our evolutionary toddlerhood? It's a safe bet they didn't involve worrying about retirement. Imagine you were a cave person roaming around the east African savannah. What kinds of concerns would occupy your waking hours? Predators would make it into your top 10 list. So would physical injury, which might very well come from those predators. In modern times, a broken leg means a trip to the doctor. In our distant past, a broken leg often meant a death sentence. The day's climate might also be a concern, the day's offering of food another. A lot of very *immediate* needs rise to the surface, needs that have nothing to do with old age.

Why immediate? Most of the survival issues we faced in our first few million years did not take hours, or even minutes, to settle. The saber-toothed tiger either ate us or we ran away from it—or a lucky few might stab it, but the whole thing was usually over in less than half a minute. Consequently, our stress responses were shaped to solve problems that lasted not for years, but for seconds. They were primarily designed to get our muscles moving us as quickly as possible, usually out of harm's way. You can see the importance of this immediate reaction by observing people who cannot mount a thorough and sudden stress response. If you had Addison's disease, for example, you would be unable to raise your blood pressure in response to severe stress, such as being attacked by a mountain lion. Your blood pressure would drop catastrophically, probably putting you into a state of debilitating shock. You would become limp. Then you would become lunch.

These days, our stresses are measured not in moments with mountain lions, but in hours, days, and sometimes months with hectic workplaces, screaming toddlers, and money problems. Our system isn't built for that. And when moderate amounts of hormone build up to large amounts, or when moderate amounts of hormone hang around too long, they become quite harmful. That's how an exquisitely tuned system can become deregulated enough to affect a dog in a metal crate—or a report card, or a performance review.

from sniffles to forgetfulness

Stress can hurt more than our brains. In the short term, acute stress can boost cardiovascular performance—the probable source of those urban legends about grandmothers lifting one end of a car to rescue their grandchildren stuck under the wheels. Over the long term, however, too much adrenaline stops regulating surges in your blood pressure. These unregulated surges create sandpaper-like rough spots on the insides of your blood vessels. The spots turn into scars, which allow sticky substances in the blood to build up there, clogging your arteries. If it happens in the blood vessels of your heart, you get a heart attack; in your brain, you get a stroke. Not surprisingly, people who experience chronic stress have an elevated risk of heart attacks and strokes.

Stress also affects our immune response. At first, the stress response helps equip your white blood cells, sending them off to fight on the body's most vulnerable fronts, such as the skin. Acute stress can even make you respond better to a flu shot. But chronic stress reverses these effects, decreasing your number of heroic white-blood-cell soldiers, stripping them of their weapons, even killing them outright. Over the long term, stress ravages parts of the immune system involved in producing antibodies. Together, these can cripple your ability to fight infection. Chronic stress also can coax your immune system to fire indiscriminately, even at targets that aren't shooting back—like your own body.

Not surprisingly, people who experience chronic stress are sick more often. A *lot* more often. One study showed that stressed individuals were three times as likely to suffer from the common cold. People were especially vulnerable to the cold-producing virus if the stressors were social in nature and lasted more than a month. They also are more likely to suffer from autoimmune disorders, such as asthma and diabetes.

To show how sensitive the immune system can be to stress, you need look no further than an experiment done with the drama department at UCLA. If you can imagine having to think all day of the most depressing things that have ever happened in your life, then acting out these feelings in front of scientists *while they are taking your blood*, you will have a pretty good idea of this Transylvanian research exercise. During the experiment, the actors practiced method acting (which asks you, if the scene calls for you to be scared, to think of something frightening, then recite your lines while plumbing those memories). One group performed using only happy memories, the other only sad. The researchers monitored their blood samples, continually looking for immune "competence." Those people who had been working with uplifting scripts all day long had healthy immune systems. Their immune cells were plentiful, happy, readily available for work. Those people who had been working with depressing scripts all day long showed something unexpected: a marked decrease in immune responsiveness. Their immune cells were not plentiful, not as robust, not as available for work. These actors were much more vulnerable to infection.

The brain is just as influenced by stress as the immune system is. The hippocampus, that fortress of human memory, is studded with cortisol receptors like cloves in a ham. This makes it *very* responsive to stress signals. If the stress is not too severe, the brain performs better. Its owner can solve problems more effectively and is more likely to retain information. There's an evolutionary reason for this. Life-threatening events are some of the most important experiences we can remember. They happened with lightning speed in the savannah, and those who could commit those experiences to memory the fastest (and recall them accurately with equal speed) were more apt to survive than those who couldn't. Indeed, research shows that memories of stressful experiences are formed almost instantaneously in the human brain, and they can be recalled very quickly during times of crises.

If the stress is too severe or too prolonged, however, stress begins to harm learning. The influence can be devastating. You can see the effects of stress on learning in everyday life. Stressed people don't do math very well. They don't process language very efficiently. They have poorer memories, both short and long forms. Stressed individuals do not generalize or adapt old pieces of information to new scenarios as well as non-stressed individuals. They can't concentrate. In almost every way it can be tested, chronic stress hurts our ability to learn. One study showed that adults with high stress levels performed 50 percent worse on certain cognitive tests than adults with low stress. Specifically, stress hurts declarative memory (things you can declare) and executive function (the type of thinking that involves problem-solving). Those, of course, are the skills needed to excel in school and business.

the villain, the hero

The biology behind this obvious assault on our intelligences can be described as a tale of two molecules, one a villain, the other a hero. The villain is the previously discussed cortisol, part of a motley crew of hormones going by the tongue-twisting name of glucocorticoids (I'll call them stress hormones). These hormones are secreted by the adrenal glands, which lie like a roof on top of your kidneys. The adrenal glands are so exquisitely responsive to neural signals, they appear to have once been a part of your brain that somehow fell off and landed in your mid-abdomen.

Stress hormones can do some truly nasty things to your brain if boatloads of the stuff are given free access to your central nervous system. That's what's going on when you experience chronic stress. Stress hormones seem to have a particular liking for cells in the hippocampus, and that's a problem, because the hippocampus is deeply involved in many aspects of human learning. Stress hormones can make cells in the hippocampus more vulnerable to other stresses. Stress hormones can disconnect neural networks, the webbing of brain cells that act like a safety deposit vault, storing your most precious memories. They can stop the hippocampus from giving birth to brand-new baby neurons. Under extreme conditions, stress hormones can even kill hippocampal cells. Quite literally, severe stress can cause brain damage in the very tissues most likely to help your children pass their SATs.

The brain seems to be aware of all this and has supplied our story not only with a villain but also with a hero. We met this champion back in the Exercise chapter. It's the Brain Derived Neurotrophic Factor. BDNF is the premier member of the powerful group of proteins called neurotrophins. BDNF in the hippocampus acts like a standing military armed with bags of Miracle Gro, keeping neurons alive and growing in the presence of hostile action. As long as there is enough BDNF around, stress hormones cannot do their damage. As I said, BDNF is a hero. How, then, does the system break down?

The problem begins when too many stress hormones hang around in the brain too long, a situation you find in chronic stress, especially of the learned helplessness variety. As wonderful as the BDNF fertilizer armies are, it is possible to overwhelm them if they are assaulted with a sufficiently strong (and sufficiently lengthy) glucocorticoid siege. Like a fortress overrun by invaders, enough stress hormones will eventually overwhelm the brain's natural defenses and wreak their havoc. In sufficient quantities, stress hormones are fully capable of turning off the gene that makes BDNF in hippocampal cells. You read that right: Not only can they overwhelm our natural defenses, but they can actually turn them *off*.

The damaging effects can be long-lasting, a fact easily observed when people experience catastrophic stress.

You might recall the bodyguard who was in the car with Princess Diana on the night of her death. To this day, he cannot remember the events several hours before or after the crash. That is a typical response to severe trauma. Its lighter cousin, forgetfulness, is quite common when the stress is perhaps less severe but more pervasive.

One of the most insidious effects of prolonged stress is that it pushes people into depression. I don't mean the type of "blues" people can experience as a normal part of daily living. Nor do I mean the type resulting from tragic circumstance, such as the death of a relative. I am talking about the kind of depression that causes as many as 800,000 people a year to attempt suicide. It is a disease every bit as organic as diabetes, and often deadlier. Chronic exposure to stress can lead you to depression's doorstep, then push you through. Depression is a deregulation of thought processes, including memory, language, quantitative reasoning, fluid intelligence, and spatial perception. The list is long and familiar. But one of its hallmarks may not be as familiar, unless you are in depression. Many people who feel depressed also feel there is no way out of their depression. They feel that life's shocks are permanent and things will never get better. Even when there is a way out—treatment is often very successful—there is no perception of it. They can no more argue their way out of a depression than they could argue their way out of a heart attack.

Clearly, stress hurts learning. Most important, however, stress hurts *people*.

a genetic buffer

In a world as complex as the brain, is the relationship between stress and learning that straightforward? For once, the answer is yes. Out-of-control stress is bad news for the brains of most people. Of course, most doesn't mean all. Like oddly placed candles in a dark room, some people illuminate corners of human behavior with unexpected clarity. They illustrate the complexity of environmental and genetic factors.

Jill was born into an inner-city home. Her father began having sex with Jill and her sister during their preschool years. Her mother was institutionalized twice because of what used to be termed "nervous breakdowns." When Jill was 7 years old, her agitated dad called a family meeting in the living room. In front of the whole clan, he put a handgun to his head, said, "You drove me to this," then blew his brains out. The mother's mental condition continued to deteriorate, and she revolved in and out of mental hospitals for years. When Mom was home, she would beat Jill. Beginning in her early teens, Jill was forced to work outside the home to help make ends meet. As Jill got older, we would have expected to see deep psychiatric scars, severe emotional damage, drugs, maybe even a pregnancy or two. Instead, Jill developed into a charming and quite popular young woman at school. She became a talented singer, an honor student, and president of her high school class. By every measure, she was emotionally well-adjusted and seemingly unscathed by the awful circumstances of her childhood.

Her story, published in a leading psychiatric journal, illustrates the unevenness of the human response to stress. Psychiatrists long have observed that some people are more tolerant of stress than others. Molecular geneticists are beginning to shed light on the reasons. Some people's genetic complement naturally buffers them against the effects of stress, even the chronic type. Scientists have isolated some of these genes. In the future, we may be able to tell stress-tolerant from stress-sensitive individuals with a simple blood test, looking for the presence of these genes.

the tipping point

How can we explain both the typical responses to stress, which can be quite debilitating, and the exceptions? For that, we turn to a senior scientist, Bruce McEwen, an elder statesman, smart, always in a suit and tie.

McEwen developed a powerful framework that allows us to understand all the various ways humans respond to stress. He gave it a name straight out of a *Star Trek* engineering manual: allostasis. Allo is from a Greek word meaning variable; stasis means a condition of balance. The idea is that there are systems that help keep the body stable by changing themselves. The stress system in the human body, and its many intricate subsystems, is one of those. The brain coordinates these body-wide changes—behavior included—in response to potential threats.

This model says that stress, left alone, is neither harmful nor toxic. Whether stress becomes damaging is the result of a complex interaction between the outside world and our physiological capacity to manage the stress. Your body's reaction to stress depends on the stress, on its length and severity, and on your body. There's a point where stress becomes toxic, and McEwen calls it the allostatic load. I know it as the first time, and only time, I ever heard my mother use profanity. I also know it as the time I got the worst grade in my academic career. We all have stories that illustrate the concrete effects of stress on real life.

As you may recall, my mother was a fourth-grade teacher. I was upstairs in my room, unbeknownst to my mother, who was upstairs in her room grading papers. She was grading one of her favorite students, a sweet, brown-haired little wisp of a girl I will call Kelly. Kelly was every teacher's dream kid: smart, socially poised, blessed with a wealth of friends. Kelly had done very well in the first half of the school year.

The second half of the school year was another story, however. My mother sensed something was very wrong the moment Kelly walked into class after Christmas break. Her eyes were mostly downcast, and within a week she had gotten into her first fight. In another week, she got her first C on an exam, and that would prove to be the high point, as her grades for the rest of the year fluttered between D's and F's. She was sent to the principal's office numerous times, and my mother, exasperated, decided to find out what caused this meltdown. She learned that Kelly's parents had decided to get a divorce over Christmas and that the family conflicts, from which the parents valiantly had insulated Kelly, had begun spilling out into the open. As things unraveled at home, things also unraveled at school. And on that snowy day, when my mother gave Kelly her third straight D in spelling, my mother also swore:

"Dammit!" she said, nearly under her breath. I froze as she shouted, "THE ABILITY OF KELLY TO DO WELL IN MY CLASS HAS NOTHING TO DO WITH MY CLASS!"

She was, of course, describing the relationship between home life and school life, a link that has frustrated teachers for a long time. One of the greatest predictors of performance in school turns out to be the emotional stability of the home.

stress in the home

I want to focus on stress in the home because it is profoundly related to kids' ability to do well in the classroom and, when they grow up, in the workforce.

Consider the all-too common case of kids witnessing their parents fighting. The simple fact is that children find unresolved marital conflict deeply disturbing. Kids cover their ears, stand motionless with clenched fists, cry, scowl, ask to leave, beg parents to stop. Study after study has shown that children—some as young as 6 months—react to adult arguments physiologically, such as with a faster heart rate and higher blood pressure. Kids of all ages who watch parents constantly fight have more stress hormones in their urine. They have more difficulty regulating their emotions, soothing themselves, focusing their attention on others. They are powerless to stop the conflict, and the loss of control is emotionally crippling. As you know, control is a powerful influence on the perception of stress.

This loss can influence many things in their lives, including their schoolwork. They are experiencing allostatic load.

I have firsthand experience with the effects of stress on grades. I was a senior in high school when my mother was diagnosed with the disease that would eventually kill her. She had come home late from a doctor's visit and was attempting to fix the family dinner. But when I found her, she was just staring at the kitchen wall. She haltingly related the terminal nature of her medical condition and then, as if that weren't enough, unloaded another bombshell. My dad, who had some prior knowledge of Mom's condition, was not handling the news very well and had decided to file for divorce. I felt as if I had just been punched in the stomach. For a few seconds I could not move. School the next day, and for the next 13 weeks, was a disaster. I don't remember much of the lectures. I only remember staring at my textbooks, thinking that this amazing woman had taught me to read and love such books, that we used to have a happy family, and that all of this was coming to an end. What she must have been feeling, much worse than I could ever fathom, she never related. Not knowing how to react, my friends soon withdrew from me even as I withdrew from them. I lost the ability to concentrate, my mind wandering back to childhood. My academic effort became a train wreck. I got the only D I would ever get in my school career, and I couldn't have cared less.

Even after all these years, it is still tough to write about that high school moment. But it easily illustrates this second, very powerful consequence of stress, underscoring with sad vengeance our Brain Rule: Stressed brains do not learn the same way as non-stressed brains. My grief at least had an end-point. Imagine growing up in an emotionally unstable home, where the stress seems never-ending. Given that stress can powerfully affect learning, one might predict that children living in high-anxiety households would not perform as well academically as kids living in more nurturing households.

That is exactly what researchers have found. Marital stress at home can negatively affect academic performance in almost every way measurable, and at nearly any age. Initial studies focused on grade-point averages over time. They reveal striking disparities of achievement between divorce and control groups. Subsequent investigations showed that even when a couple stays together, children living in emotionally unstable homes get lower grades. (Careful subsequent investigations showed that it was the presence of overt conflict, not divorce, that predicted grade failure.) They also do worse on standardized math and reading tests.

The stronger the degree of conflict, the greater the effect on performance. Teachers typically report that children from disrupted homes rate lower in both aptitude and intelligence. Such children are three times as likely to be expelled from school or to become pregnant as teenagers, and five times as likely to live in poverty. As social activist Barbara Whitehead put it, writing for the *Atlantic Monthly*: "Teachers find many children emotionally distracted, so upset and preoccupied by the explosive drama of their own family lives that they are unable to concentrate on such mundane matters as multiplication tables."

Physical health deteriorates; absenteeism and truancy increase. The absenteeism may occur because stress is depleting the immune system, which increases the risk of infection. Though the evidence is not as conclusive, a growing body of data suggests that children living in hostile environments are at greater risk for certain psychiatric disorders, such as depression and anxiety disorders. Such disorders can wreak havoc on cognitive processes important to successful academic performance. As children grow up, the effects of childhood stress can stay with them. Indeed, performance can take a negative hit regardless of one's age, even if you were a previously high functioning and much admired employee, like Lisa Nowak.

stress at work

You may have heard of Lisa Nowak. She is a lethal combat pilot, decorated electronics warfare specialist, pretty, smart. The government spent millions of dollars training her to be an astronaut. She was also a mother with two kids on the verge of divorcing her husband one month before her biggest professional assignment: mission control specialist for a shuttle mission. Talk about built-up stress. She put some weapons in her automobile, grabbed a disguise, even packed up a bunch of adult diapers so she didn't have to stop to use a bathroom. She then drove virtually nonstop from Orlando to Houston, allegedly to kidnap her target, a woman she thought was a threat to a fellow astronaut to whom she had taken a fancy. Instead of serving as the lead for one of America's most technically challenging jobs, this highly skilled engineer is awaiting trial on attempted kidnapping and burglary charges. She will probably never fly again, which makes this sad story nearly heartbreaking. It also makes the money spent on her training a colossal waste. But those few million dollars are miniscule compared with the cost of stress on the workplace as a whole.

Stress attacks the immune system, increasing employees' chances of getting sick. Stress elevates blood pressure, increasing the risk of heart attack, stroke, and autoimmune diseases. That directly affects health-care and pension costs. Stress is behind more than half of the 550 million working days lost each year because of absenteeism. Stressed employees tend to avoid coming to work at the slightest excuse, and they often show up late. Yet executives often give stress the shortest shrift. The Centers for Disease Control and Prevention asserts that a full 80 percent of our medical expenditures are now stress-related. In a work force where 77 percent report being burned out, this translates into a lot of cortisol, a lot of missed meetings, and a lot of trips to the doctor. That's not all. Prolonged stress can cause depression, which alters the ability to think—a direct assault on a corporation's intellectual capital. This injury to business productivity is threefold.

First, depression hobbles the brain's natural improvisatory instincts the way arthritis hobbles a dancer. Fluid intelligence, problem-solving abilities (including quantitative reasoning), and memory formation are deeply affected by depression. The result is an erosion of innovation and creativity, just as biochemically real as if we were talking about joints and muscles. In a knowledge-based economy where intellectual dexterity is often *the* key to survival, that's bad news for competitiveness, shareholder value, and the bottom line. In fact, the cost of depression to the work force in 1990 was estimated to be $53 billion. The loss of productivity contributed the most, about $33 billion of the total.

Second, those same people who have lost their creativity incur more health-care expenses. Thus, not only does stress reduce the contributions valuable employees can make, but those same employees begin to cannibalize their company's internal resources. And it's not just mental-health expenditures. Depressed individuals are at increased risk for a number of other diseases.

Third, people who burn out are often fired, if they don't leave on their own. Turnover further disrupts productivity, plus sets off a costly recruiting and training effort. The ugly truth is that any assault on human brain cells is an assault on competitiveness. The final tab? Statistical analyses from many studies form the same dismal picture. Stress causes companies to lose between $200 billion and $300 billion a year—as much as $75 billion of red ink a quarter.

Three things matter in determining whether a workplace is stressful: the type of stress, a balance between occupational stimulation and boredom, and the condition of the employee's home life. Business professionals have spent a long time studying what types of stress make people less productive and, not surprisingly, have arrived at the same conclusion that Marty Seligman's German shepherds did: Control is critical. The perfect storm of occupational stress appears to be a combination of two malignant facts: a) a great deal is expected of you and b) you have no control over whether you will perform well. Sounds like a formula for learned helplessness to me.

On the positive side, restoration of control can return groups to productivity. In one instance, a for-profit consented to be studied after agreeing to institute a control-based stress management program. At the end of two years, the unit had saved almost $150,000 in workers' compensation costs alone. The cost of deploying the stress management program? About $6,000. And just 16 hours of the program reduced toxic blood pressure levels for employees diagnosed with hypertension.

Control isn't the only factor in productivity. Employees on an assembly line, doing the same tired thing day after day, are certainly in control of their work processes. But the tedium can be a source of brain-numbing stress. What spices things up? Studies show that a certain amount of uncertainty can be good for productivity, especially for bright, motivated employees. What they need is a balance between controllability and uncontrollability. Slight feelings of uncertainty may cause them to deploy unique problem-solving strategies.

The third characteristic, if you are a manager, is none of your business. I am talking about the effects of family life on business life. There's no such thing as a firewall between personal issues and work productivity. That's because we can't have two brains we can interchange depending upon whether we are in our office or in our bedroom. Stress in the workplace affects family life, causing more stress in the family. More stress in the family causes more stress at work, which in turn gets brought home again. It's a deadly, self-feeding spiral, and researchers call it "work-family conflict." So you may have the most wonderful feelings about autonomy at work, and you may have tremendous problem-solving opportunities with your colleagues. But if your home life is a wreck, you can still suffer the negative effects of stress, and so can your employer.

Whether we look at school performance or job performance, we keep running into the profound influence of the emotional stability of the home. Is there anything we can do about something so fundamentally personal, given that its influence can be so terribly public? The answer, surprisingly, may be yes.

marriage intervention

Famed marriage researcher John Gottman can predict the future of a relationship within three minutes of interacting with the couple. His ability to accurately forecast marital success or failure is close to 90 percent. His track record is confirmed by peer-reviewed publications. He may very well hold the future of the American education and business sectors in his hands.

How is he so successful? After years of careful observation, Gottman isolated specific marital behaviors—both positive and negative—that hold most of the predictive power. But this research was ultimately unsatisfying to a man like Gottman, akin to telling someone they have a life-threatening illness but not being able to cure them. And so the next step in his research was to try to harness some of that predictive knowledge to give a couple a better future. Gottman devised a marriage intervention strategy based on his decades of research. It focuses on improving the behaviors proven to predict marital success and eliminating the ones proven to predict failure. Even in its most modest forms, his intervention drops divorce rates by nearly 50 percent.

What do his interventions actually do? They drop both the frequency and severity of hostile interactions between husband and wife. This return to civility has many positive side effects besides marital reconstruction, especially if the couple has kids. The link is direct. These days, Gottman says, he can predict the quality of a relationship not only by examining the stress responses of the parents but also by taking a urine sample of their children.

That last statement deserves some unpacking. Gottman's marriage research invariably put him in touch with couples who were starting families. When these marriages started their transition to parenthood, Gottman noticed that the couple's hostile interactions skyrocketed. There were many causes, ranging from chronic sleep deprivation to the increased demands of a helpless new family member (little ones typically require that an adult satisfy some demand of theirs about 3 times a minute). By the time the baby was 1 year old, marital satisfaction had plummeted 70 percent. At that same point, the risk for maternal depression went from 25 percent to a whopping 62 percent. The couples' risk for divorce increased, which meant American babies often were born into a turbulent emotional world.

That single observation gave Gottman and fellow researcher Alyson Shapiro an idea. What if he deployed his proven marital intervention strategies to married couples while the wife was pregnant? Before the hostility floodgates opened up? Before the depression rates went through the roof? Statistically, he already knew the marriage would significantly improve. The big question concerned the kids. What would an emotionally stabilized environment do to the baby's developing nervous system? He decided to find out.

The research investigation, deployed over several years, was called Bringing Baby Home. It consisted of exposing expectant couples to the marital interventions whether their marriages were in trouble or not, and then assessing the development of the child. Gottman and Shapiro uncovered a gold mine of information. They found that babies raised in the intervention households didn't look anything like the babies raised in the controls. Their nervous systems didn't develop the same way. Their behaviors weren't in the same emotional universe. Children in the intervention groups didn't cry as much. They had stronger attention-shifting behaviors and they responded to external stressors in remarkably stable ways. Physiologically, the intervention babies showed all the cardinal signs of healthy emotional regulation, while the controls showed all the signs of unhealthy, disorganized nervous systems. The differences were remarkable and revealed something hopeful and filled with common sense. By stabilizing the parents, Gottman and Shapiro were able to change not only the marriage; they also were able to change the child.

I think Gottman's findings can change the world, starting with report cards and performance evaluations.

ideas

What people do in their private life is their own business, of course. Unfortunately, what people do in their private life often affects the public. Consider the criminal history of a fellow who had recently moved from Texas to a city in the Pacific Northwest. He absolutely *hated* his new home and decided to leave. Stealing the car of a neighbor (for the second time that month), he drove several miles to the airport and ditched the car. He then found a way to fool both the security officials and the gate managers and hopped a free ride back to Texas. He accomplished this feat a few months shy of his 10th birthday. Not surprisingly, this boy comes from a troubled home. This is a fairly recent event, but if something isn't done soon, the private issue of raising this child soon will become a very public problem. And he is hardly alone. How can we capture our Brain Rule, that stressed brains learn differently from non-stressed brains, and change the way we educate, parent, and do business? I have thought a lot about that.

Teach parents first

The current education system starts in first grade, typically around age 6. The curriculum is a little writing, a little reading, a little math. The teacher is often a complete stranger. And there is something important missing. The stability of the home is completely ignored, even though it is one of the greatest predictors of future success at school. But what if we took its influence seriously?

My idea envisions an educational system where the first students are not the children. The first students are the parents.

The curriculum? How to create a stable home life, using Gottman's powerful, baby-nervous-system changing protocols. The intervention could even start in a maternity ward, offered by a hospital (like a Lamaze class, which takes just about as much time). There would be a unique partnership between the health system and the education system. *And it makes education, from the beginning of a child's life, a family affair.*

First grade would begin a week after birth. The amazing cognitive abilities of infants, from language acquisition to the powerful need for luxurious amounts of active playtime, are fully unleashed in a curriculum designed just for them. (This is *not* a call to implement products in the strange industry that seeks to turn babies into Einsteins in the first year of life. Most of those products have not been tested, and some have been shown to be harmful to learning. My idea envisions a mature, rigorously tested pedagogy that does not yet exist—one more reason for educators and brain scientists to work together.) Along with this, parents would take an occasional series of marital refresher courses, just to ensure the stability of the home. Can you imagine what a child might look like academically after years of thriving in an emotionally stable environment? The child flourishes in this fantasy.

No hospitals or schools currently offer these interventions to the future students of America, and there is no formal curriculum for harnessing the cognitive horsepower of the under-solid-food crowd. But it could be developed and tested, beginning right this minute. The best shot would come from collaborative experiments between brain scientists and education scientists. All one needs is a cooperative educational will, and maybe a sense of adventure.

Free family counseling, child care

Historically, people have done their best work—sometimes world-changing work—in their first few years after joining the work force. In the field of economics, most Nobel Prize-winning research is done in the first 10 years of the recipient's career. Albert Einstein published most of his creative ideas at the ripe old age of 26. It's no wonder that companies want to recruit young intellectual talent.

The problem in today's economy is that people are typically starting a family at the very time they are also supposed to be doing their best work. They are trying to be productive at some of the most stressful times of their lives. What if companies took this unhappy collision of life events seriously? They could offer Gottman's intervention as a benefit for every newly married, or newly pregnant, employee. Would that reverse the negative flow of family stress that normally enters the workplace at this time in a person's life? Such an intervention might enhance productivity and perhaps even generate grateful, loyal employees.

Businesses also risk losing their best and brightest at this time, as talented people are forced to make a terrible decision between career and family. The decision is especially hard on women. In the 21st century, we have invented two economic classifications: the child-free class (people with no kids or without primary responsibility for them) and the child-bound class (people who act as a main caregiver). From a gender perspective, these categories have very little symmetry. According to Claudia Goldin, Henry Lee Professor of Economics at Harvard, women are overrepresented in the child-bound category by nearly 9 to 1.

What if talented people didn't have to choose between career and family? What if businesses offered onsite child care just so they could retain employees at the very time they are most likely to be valuable? This obviously would affect women the most, which means businesses immediately achieve more gender balance. Would such an offering so affect productivity that the costs of providing child care become offset by the gains? That's a great research question. Not only might businesses create more stable employees in the current generation, they might be raising far healthier children for work in the next.

Power to the workers

Plenty of books discuss how to manage stress; some are confusing, others extraordinarily insightful. The good ones all say one thing in common: The biggest part of successful stress management involves getting control back into your life. This means that a manager or human-resources professional has a powerful predictive insight at his or her disposal. To detect stress-related problems, one might simply examine the situations where an employee feels the most helpless. Questionnaires based on Jeansok Kim's and David Diamond's three-pronged definition of stress could be developed that routinely assess not the broad perception of aversion, but the narrower issue of powerlessness. The next step would be to change the situation.

These are only a few of the possibilities that could be realized if brain scientists and business professionals ever collaborated on the biology of stress in the work force. It is possible their findings would change the absentee rate of their employees, cut down on the number of trips to the doctor, and reduce their insurance overhead. As well as money saved, a great deal of creativity may be engendered simply by routinely giving employees a way out—not from their jobs but from the stress they experience in them.

It's no coincidence that stress researchers, education scientists, and business professionals come to similar conclusions about stress and people. What's astonishing is that we have known most of the salient points since Marty Seligman stopped shocking his dogs in the mid-1970s. It is time we made productive use of that horrible line of research.

<u>Summary</u>

Rule #8
Stressed brains
don't learn the same way.

* Your body's defense system—the release of adrenaline and cortisol—is built for an immediate response to a serious but passing danger, such as a saber-toothed tiger. Chronic stress, such as hostility at home, dangerously deregulates a system built only to deal with short-term responses.

* Under chronic stress, adrenaline creates scars in your blood vessels that can cause a heart attack or stroke, and cortisol damages the cells of the hippocampus, crippling your ability to learn and remember.

* Individually, the worst kind of stress is the feeling that you have no control over the problem—you are helpless.

* Emotional stress has huge impacts across society, on children's ability to learn in school and on employees' productivity at work.

Get more at www.brainrules.net

sensory integration

Rule #9

Stimulate more of the senses.

EVERY TIME TIM SEES the letter "E," he also sees the color red. He describes the color change as if suddenly forced to look at the world through red-tinted glasses. When Tim looks away from the letter "E," his world returns to normal, until he encounters the letter "O." Then the world turns blue. For Tim, reading a book is like living in a disco. For a long time, Tim thought this happened to everyone. When he discovered this happened to *no* one—at least not in his immediate circle—he began to suspect he was crazy. Neither impression was correct, of course. Tim is suffering—if that's the right word—from a brain condition called synesthesia. Though experienced by as many as 1 in 2,000 people (some think 1 in 200), it is a behavior about which scientists know next to nothing. At first blush, there appears to be a short-circuiting between the processing of various sensory inputs. If scientists can nail down what happens when sensory processing goes wrong, they may gain more understanding about what happens when it goes right. So, synesthesia intrigues scientists interested in how the brain processes the world's many senses. The effect that this has on learning forms the heart of our Brain Rule: Stimulate more of the senses at the same time.

saturday night fever

That you can detect anything has always seemed like a minor miracle to me. On one hand, the inside of your head is a darkened, silent place, lonely as a cave. On the other hand, your head crackles with the perceptions of the whole world, sight, sound, taste, smell, touch, energetic as a frat party. How could this be? For a long time, nobody could figure it out. The Greeks didn't think the brain did much of anything. It just sat there like an inert pile of clay (indeed, it does not generate enough electricity to prick your finger). Aristotle thought the heart held all the action, pumping out rich, red blood 24 hours a day. The heart, he reasoned, harbored the "vital flame of life," a fire producing enough heat to give the brain a job description: to act as a cooling device (he thought the lungs helped out, too). Perhaps taking a cue from our Macedonian mentor, we still use the word "heart" to describe many aspects of mental life.

How does the brain, brooding in its isolated bony chambers, perceive the world? Consider this example: It is Friday night at a New York club. The dance beat dominates, both annoying and hypnotic, felt more than heard. Laser lights shoot across the room. Bodies move. The smells of alcohol, fried food, and illegal smoking mix in the atmosphere like a second sound track. In the corner, a jilted lover is crying. There is so much information in the room, you are beginning to get a headache, so you step out for a breath of fresh air. The jilted lover follows you.

Snapshots like these illustrate the incredible amount of sensory information your brain must process simultaneously. External physical inputs and internal emotional inputs all are presented to your brain in a never-ending fire hose of sensations. Dance clubs may seem the extreme. Yet there may be no more information there than what you'd normally experience the next morning on the streets of Manhattan. Faithfully, your brain perceives the screech of the taxis, the pretzels for sale, the crosswalk signal, and the people brushing past, just as it could hear the pounding beat and smell the cigarettes last night. You are a wonder. And we in brain-science land are only beginning to figure out how you do it.

Scientists often point to an experience called the McGurk effect to illustrate sensory integration. Suppose researchers showed you a video of a person saying the surprisingly ugly syllable "ga." Unbeknownst to you, the scientists had turned off the sound of the original video and had dubbed the sound "ba" onto it. When the scientist asks you to listen to the video with your eyes closed, you hear "ba" just fine. But if you open your eyes, your brain suddenly encounters the shape of the lips saying "ga" while your ears are still hearing "ba." The brain has no idea what to do with this contradiction. So it makes something up. If you are like most people, what you actually will hear when your eyes open is the syllable "da." This is the brain's compromise between what you hear and what you see—its need to attempt integration.

But you don't have to be in a laboratory to show this. You can just go to a movie. The actors you see speaking to each other on screen are not really speaking to each other at all. Their voices emanate from speakers cleverly placed around the room: some behind you, some beside you; none centered on the actors' mouths. Even so, you believe the voices are coming from those mouths. Your eye observes lips moving in tandem with the words your ears are hearing, and the brain combines the experience to trick you into believing the dialogue comes from the screen. Together, these senses create the perception of someone speaking in front of you, when actually nobody is speaking in front of you.

how the senses integrate

Analyses like these have led scientists to propose a series of theories about how the senses integrate. On one end of this large continuum are ideas that remind me of the British armies during the Revolutionary War. On the other end are ideas that remind me of how the Americans fought them. The British, steeped in the traditions of large European land wars, had lots of central planning. The field office gathered information from leaders on the battleground and then issued its commands. The Americans, steeped in the traditions of nothing, used guerrilla tactics: on-the-ground analysis and decision making prior to consultation with a central command.

Take the sound of a single gunshot over a green field during that war. In the British model of this experience, our senses function separately, sending their information into the brain's central command, its sophisticated perception centers. Only in these centers does the brain combine the sensory inputs into a cohesive perception of the environment. The ears hear the rifle and generate a complete auditory report of what just occurred. The eyes see the smoke from the gun arising from the turf and process the information separately, generating a visual report of the event. The nose, smelling gunpowder, does the same thing. They each send their data to central command. There, the inputs are bound together, a cohesive perception is created, and the brain lets the soldier in on what he just experienced. The processes can be divided into three steps:

STEP 1: SENSATION

This is where we capture the energies from our environment pushing themselves into our orifices and rubbing against our skin. The effort involves converting this external information into a brain-friendly electrical language.

STEP 2: ROUTING

Once the information is successfully translated into head-speak, it is sent off to appropriate regions of the brain for further processing.

The signals for vision, hearing, touch, taste and smell all have separate, specialized places where this processing occurs. A region called the thalamus, that well-connected, egg-shaped structure in the middle of your "second brain," helps supervise most of this shuttling.

STEP 3: PERCEPTION

The various senses start merging their information. These integrated signals are sent to increasingly complex areas of the brain (actually called higher regions), and we begin to perceive what our senses have given us. As we shall see shortly, this final step has both bottom-up and top-down features.

The American model puts things very differently. Here the senses work together from the very beginning, consulting and influencing one another quite early in the process. As the ear and eye simultaneously pick up gunshot and smoke, the two impressions immediately confer with each other. They perceive that the events are occurring in tandem, without conferencing with any higher authority. The picture of a rifle firing over an open field emerges in the observer's brain. The steps are still sensation, routing, and perception. But at each step, add "the signals begin immediately communicating, influencing subsequent rounds of signal processing." The last stage, perception, is not where the integration begins. The last step is where the integration culminates.

Which model is correct? The data are edging in the direction of the second model, but the truth is that nobody knows how it works. There are tantalizing suggestions that the senses actually help one another, and in a precisely coordinated fashion. This chapter is mostly interested in what happens after sensation and routing—after we achieve perception.

bottoms up, tops down

We can see how important this last step is by looking at what happens when it breaks down. Oliver Sacks reports on a patient he calls Dr. Richard who had lost various perceptual processing abilities. There wasn't anything wrong with Dr. Richard's vision. He just couldn't always make sense of what he saw. When a friend walked into the room and sat down on a chair, he did not always perceive the person's various body parts as belonging to the same body. Only when the person got up out of the chair would he suddenly recognize them as possessed by one person. If Dr. Richard looked at a photograph of people at a football stadium, he would identify the same colors of different people's wardrobes as belonging "together" in some fashion. He could not see such commonalities as belonging to separate people. Most interesting, he could not always perceive multisensory stimuli as belonging to the same experience. This could be observed when Dr. Richard tried to watch someone speaking. He sometimes could not make a connection between the motion of the speaker's lips and the speech. They would be out of sync; he sometimes reported the experience as if watching a "badly dubbed foreign movie."

Given the survival advantage to seeing the world as a whole, scientists have been deeply concerned with the binding problem. They ask: Once the thalamus has done its distribution duties, what happens next? The information, dissected into sensory-size pieces and flung widely across the brain's landscape, needs to be reassembled (something Dr. Richard was not very good at). Where and how does information from different senses begin to merge in the brain?

The where is easier than the how. We know that most of the sophisticated stuff occurs in regions known as association cortices. Association cortices are specialized areas that exist throughout the brain, including the parietal, temporal, and frontal lobes. They are not exactly sensory regions, and they are not exactly motor regions, but they are exactly bridges between them (hence the name *association*). Scientists think these regions use both bottom-up and top-down processes to achieve perception. As the sensory signals ascend through higher and higher orders of neural processing, these processes kick in. Here's an example.

Author W. Somerset Maugham once said: "There are only three rules for writing a novel. Unfortunately, nobody knows what they are." After your eyes read that sentence and the thalamus has spattered various aspects of the sentence all over the inside of your skull, bottom-up processors go to work. The visual system (which we will say more about in the Vision chapter) is a classic bottom-up processor. What happens? Feature detectors—which work like auditors in an accounting firm—greet the sentence's visual stimuli. The auditors inspect every structural element in each letter of every word in Maugham's quote. They write a report, a visual conception of letters and words. An upside-down arch becomes the letter "U." Two straight lines at right angles become the letter "T." Combinations of straight lines and curves become the word "three." Written information has a lot of visual features in it, and this report takes a great deal of effort and time to organize. It is one of the reasons that reading is a relatively slow way to put information into the brain.

Next comes top-down processing. This can be likened to a board of directors reading the auditor's report and then reacting to it. Many comments are made. Sections are analyzed in light of pre-existing knowledge. The board in your brain has heard of the word "three" before, for example, and it has been familiar with the concept of rules since you were familiar with anything. Some board members even have heard of W. Somerset Maugham before, and they recall to your consciousness a movie called *Of Human Bondage*, which you saw in a film history course. Information is added to the data stream or subtracted from the data stream. The brain can even alter the data stream if it so chooses. And it so chooses a lot.

Such interpretive activity is the domain of top-down processing. At this point, the brain generously lets you in on the fact that you are perceiving something. Given that people have unique previous experiences, they bring different interpretations to their top-down analyses. Thus, two people can see the same input and come away with vastly different perceptions. It is a sobering thought. There is no guarantee that your brain will perceive the world accurately, even if other parts of your body can.

So, life is filled with the complex qualities of sounds, visual images, shapes, textures, tastes, and odors, and the brain seeks to simplify this world by adding more confusion. This requires large groups of receptors, each one in charge of a particular sensory attribute, to act simultaneously. For us to savor the richness and diversity of perception, the central nervous system must integrate the activity of entire sensory populations. It does this by pushing electrical signals through an almost bewildering thicket of ever more complex, higher neural assemblies. Finally, you perceive something.

survival by teamwork

There are many types of synesthesia—more than 50, according to one paper. One of the strangest illustrates that even when the brain's wiring gets confused, the senses still work together. There are some people who see a word and immediately experience a taste on their tongue. This isn't the typical mouth-watering response, such as imagining the taste of a candy bar after hearing the word "chocolate." This is like seeing the word "sky" in a novel and suddenly tasting a sour lemon in your mouth. A clever experiment showed that even when the synesthete could not recall the exact word, he or she could still get the taste, as long as there was some generalized description of the missing word. Data like these illustrate that sensory processes are wired to work together. Thus, the heart of the Brain Rule: Stimulate more of the senses.

The evolutionary rationale for this observation is simple: Our East African crib did not unveil its sensory information one sense at a time during our development. It did not possess *only* visual stimuli, like a silent movie, and then suddenly acquire an audio track a few million years later, and then, later, smells and odors and textures. By the time we came out of the trees, our ancestors were encountering a multisensory world and were already champions at experiencing it.

Some interesting experiments support these ideas. Several years ago, scientists were able to peer in on the brain using fMRI technology. They played a trick on their subjects: They showed a video of someone speaking but completely turned off the sound. When the researchers examined what the brain was doing, they found that the area responsible for processing the sound, the auditory cortex, was stimulated as if the person actually were hearing sound. If the person was presented with a person simply "making faces," the auditory cortex was silent. It had to be a visual input *related* to sound. Clearly, visual inputs influence auditory inputs, even with the sound turned off.

In another experiment at about the same time, researchers showed short flashes of light near the subjects' hands, which were rigged with a tactile stimulator. Sometimes researchers would turn on the stimulator while the flash of light was occurring, sometimes not. No matter how many times they did this, the visual portion of the brain always lighted up the strongest when the tactile response was paired with it. They could literally get a boost in the visual system by introducing touch. This effect is called multimodal reinforcement.

Multiple senses affect our ability to detect stimuli, too. Most people, for example, have a very hard time seeing a flickering light if the intensity of the light is gradually decreased. Researchers decided to test that threshold by precisely coordinating a short burst of sound with the light flickering off. The presence of sound actually changed the threshold. The subjects found that they could see the light way beyond their normal threshold if sound was part of the experience.

These data show off the brain's powerful integrative instincts. Knowing that the brain cut its developmental teeth in an overwhelmingly multisensory environment, you might hypothesize that its learning abilities are increasingly optimized the more multisensory the environment becomes. You might further hypothesize that the opposite is true: Learning is less effective in a unisensory environment. That is exactly what you find, and it leads to direct implications for education and business.

the learning link

Cognitive psychologist Richard Mayer probably has done more than anybody else to explore the link between multimedia exposure and learning. He sports a 10-megawatt smile, and his head looks exactly like an egg (albeit a very clever egg). His experiments are just as smooth: Divide the room into three groups. One group gets information delivered via one sense (say, hearing), another the same information from another sense (say, sight), and the third group the same information delivered as a combination of the first two senses.

The groups in the multisensory environments always do better than the groups in the unisensory environments. They have more accurate recall. Their recall has better resolution and lasts longer, evident even 20 *years* later. Problem-solving improves. In one study, the group given multisensory presentations generated more than 50 percent more creative solutions on a problem-solving test than students who saw unisensory presentations. In another study, the improvement was more than 75 percent!

The benefits of multisensory inputs are physical as well. Our muscles react more quickly, our threshold for detecting stimuli improves, and our eyes react to visual stimuli more quickly. It's not just combinations of sight and sound. When touch is combined with visual information, recognition learning leaps forward by almost 30 percent, compared with touch alone. These improvements are greater than what you'd predict by simply adding up the unisensory data. This is sometimes called supra-additive integration. In other words, the positive contributions of multisensory presentations are greater than the sum of their parts. Simply put, multisensory presentations are the way to go.

Many explanations have been put forth to explain these

consistent findings, and most involve working memory. You might recall from Chapter 5 that working memory, formerly called short-term memory, is a complex work space that allows the learner to hold information for a short period of time. You might also recall its importance to the classroom and to business. What goes on in the volatile world of working memory deeply affects whether something that is taught will also be learned.

All explanations about multisensory learning also deal with a counter-intuitive property lurking at its mechanistic core: Extra information given at the moment of learning makes learning better. It's like saying that if you carry two heavy backpacks on a hike instead of one, you will accomplish your journey more quickly. This is the "elaborative" processing that we saw in the chapter on short-term memory. Stated formally: It is the extra cognitive processing of information that helps the learner to integrate the new material with prior information. Multisensory experiences are, of course, more elaborate. Is that why they work? Richard Mayer thinks so. And so do other scientists, looking mostly at recognition and recall.

One more example of synesthesia supports this, too. Remember Solomon Shereshevskii's amazing mental abilities? He could hear a list of 70 words once, repeat the list back without error (forward or backward), and then reproduce the same list, again without error, 15 years later. Shereshevskii had multiple categories of (dis)ability. He felt that some colors were warm or cool, which is common. But he also thought the number 1 was a proud, well-built man, and that the number 6 was a man with a swollen foot, which was not common. Some of his imaging was nearly hallucinatory. He related: "One time I went to buy some ice cream ... I walked over to the vendor and asked her what kind of ice cream she had. 'Fruit ice cream,' she said. But she answered in such a tone that a whole pile of coals, of black cinders, came bursting out of her mouth, and I couldn't bring myself to buy any ice cream after she had answered that way."

Shereshevskii clearly is in his own mental universe, but he

illustrates a more general principle. Synesthetes almost universally respond to the question "What good does this extra information do?" with an immediate and hearty "It helps you remember." Given such unanimity, researchers have wondered for years if there is a relationship between synesthesia and advanced mental ability.

There is. Synesthetes usually display unusually advanced memory ability—photographic memory, in some cases. Most synesthetes report the odd experiences as highly pleasurable, which may, by virtue of dopamine, aid in memory formation.

rules for the rest of us

Over the decades, Mayer has isolated a number of rules for multimedia presentation, linking what we know about working memory with his own empirical findings on how multimedia exposure affects human learning. Here are five of them in summary form:

1) *Multimedia principle:* Students learn better from words and pictures than from words alone.

2) *Temporal contiguity principle:* Students learn better when corresponding words and pictures are presented simultaneously rather than successively.

3) *Spatial contiguity principle:* Students learn better when corresponding words and pictures are presented near to each other rather than far from each on the page or screen.

4) *Coherence principle:* Students learn better when extraneous material is excluded rather than included.

5) *Modality principle:* Students learn better from animation and narration than from animation and on-screen text.

Though wonderfully empirical, these principles are relevant only to combinations of two senses: hearing and vision. We have three other senses also capable of contributing to the educational environment. Beginning with the story of a talented combat veteran, let's explore what happens if we add just one more: smell.

nosing it out

I once heard a story about a man who washed out of medical school because of his nose. To understand his story, you have to know something about the smell of surgery. And you have to have killed somebody. Surgery can be a smelly experience. When you cut somebody's body, you invariably cut their blood vessels. To keep the blood from interfering with the operation, surgeons use a cauterizing tool, hot as a soldering iron. It's applied directly to the wound, burning it shut, filling the room with the acrid smell of smoldering flesh. Combat can smell the same way. And the medical student in question was a Vietnam vet with heavy combat experience. He didn't seem to suffer any aversive effects when he came home. He had no post-traumatic stress disorder, and he became a high-functioning undergraduate, eventually accepted into medical school. But then the former soldier started his first surgery rotation. Entering the surgical suite, he promptly smelled the burning flesh from the cauterizer. The smell brought back to mind the immediate memory of an enemy combatant he had shot in the face, point blank, an experience he had suppressed for years. The memory literally doubled him over. He ran out of the room crying, the dying enemy's strange gurgling sounds ringing in his ears, the noises of the evacuation helicopters in the distance. All that day, he relived the experience; later that night, he began to recall in succession other equally terrible events. He resigned from the program the next week.

This story illustrates something scientists have known for years: Smell can evoke memory. It's called the Proust effect. Marcel Proust, the French author of the profoundly moving book *Remembrance of Things Past*, talked freely 100 years ago about smells and their ability to elicit long-lost memories. Typical experiments have investigated the unusual ability of a smell to enhance retrieval. Two groups of people might be assigned to see a movie together, for example, and then told to report to the lab for a memory test. The control group goes into an unmanipulated room and simply takes the test. The experimental group takes the test in a room flooded with the smell of popcorn. The results are then compared, scoring for number of events recalled, accuracy of events recalled, specific characteristics, and so on. The results of the test can be astonishing. Some researchers report that smell-exposed experimental groups can accurately retrieve twice as many memories as the controls. Others report a 20 percent improvement, still others only a 10 percent.

One way to react to these data is to say, "Wow." Another is to ask, "Why the disparity in results?" One big reason is that the results depend on the type of memory being assessed and the methodology employed to obtain them. For example, researchers have found that certain types of memory are exquisitely sensitive to smells and other types nearly impenetrable. Odors appear to do their finest work when subjects are asked to retrieve the emotional details of a memory, as our medical student experienced, or to retrieve autobiographical memories. You get the best results if the smells are congruent. A movie test in which the smell of gasoline is pumped into the experimental room does not yield the same positive memory-retrieval results as the smell of popcorn does.

Odors are not so good at retrieving declarative memory. You can get smell to boost declarative scores, but only if the test subjects are emotionally aroused—usually, that means stressed—before the experiment begins. (For some reason, showing a film of young Australian aboriginal males being circumcised is a favorite way to do this). Recent tests, however, show that smell can improve declarative memory recall during sleep, a subject we will take up in a moment. Is there a reason why the Proust effect exists—why smell evokes memory? There might be, but to understand it, we have to know a little bit about how smell is processed in the brain.

Right between the eyes lies a patch of neurons about the size of a large postage stamp. This patch is called the olfactory region. The outer surface of this region, the one closest to the air in the nose, is the olfactory epithelium. When we sniff, odor molecules enter the nose chamber and collide with nerves there. This in itself is amazing, given that the chamber is always covered by a thick layer of snot. Somehow these persistent biochemicals penetrate the mucus and brush against little quill-like protein receptors that stud the nerves in the olfactory epithelium. The receptors can recognize a large number of smell-evoking molecules. When that happens, the neurons begin to fire excitedly, and you are well on your way to smelling something. The rest of the journey occurs in the brain. The now occupied nerves of the olfactory epithelium chat like teenagers on a cell phone to a group of nerves lying directly above them, in the olfactory bulb. These nerves help sort out the signals sent to it by the epithelium.

Here comes the interesting part of the story. Every other sensory system, at this point, must send a signal to the thalamus and ask permission to connect to the rest of the brain—including the higher levels where perception occurs. Not the nerves carrying information about smell. Like an important head of state in a motorcade, smell signals bypass the thalamus and go right to their brainy destinations, no meddling middle-man required.

One of those destinations is the amygdala, and it is at this point that the Proust effect begins to make some sense. As you recall, the amygdala supervises not only the formation of emotional experiences but also the *memory* of emotional experiences. Because smell directly stimulates the amygdala, smell directly stimulates emotions. Smell signals also head through the piriform cortex to the orbitofrontal cortex, a part of your brain just above and behind your eyes and deeply involved in decision making. So smell plays a role in decision making. It is almost as if the odor is saying, "My signal is so important, I am going to give you a memorable emotion. What are you going to do about it?"

Smell signals appear to be in a real hurry to take these shortcuts, so much so that olfactory receptor cells aren't even guarded by a protective barrier. This is different from most other sensory receptor cells in the human body. Visual receptor neurons in the retina are protected by the cornea, for example. Receptor neurons that allow hearing in our ears are protected by the eardrum. The only things protecting receptor neurons for smell are boogers. Otherwise, they are directly exposed to the air.

ideas

There is no question that multiple cues, dished up via different senses, enhance learning. They speed up responses, increase accuracy, improve stimulation detection, and enrich encoding at the moment of learning. Yet we still aren't harnessing these benefits on a regular basis in our classrooms and boardrooms. Here are a couple of ideas that come to mind.

Multisensory school lessons

As we learned in the Attention chapter, the opening moments of a lecture are cognitive hallowed ground. It is the one time teachers automatically have more student minds paying attention to them. If presentations during that critical time were multisensory, overall retention might increase. We discovered in the Memory chapters that repeating information in timed intervals helps stabilize memory. What if we introduced information as a multisensory experience, and then repeated not only the information but also one of the modes of presentation? The first re-exposure might be presented visually, for example; the next, auditorially; the third, kinesthetically. Would that encoding-rich schedule increase retention in real-world environments, boosting the already robust influence of repetition?

And let's not continue to neglect our other senses. We saw that touch and smell are capable of making powerful contributions to the learning process. What if we began to think seriously about how to adopt them into the classroom, perhaps in combination with more traditional learning presentations? Would we capture their boosting effects, too?

One study showed that a combination of smell and sleep improved declarative-memory consolidation. The delightful experiment used a card game my sons and I play on a regular basis. The game involves a specialized 52-card deck we purchased at a museum, resplendent with 26 pairs of animals. We turn all of the cards face down, then start selecting two cards to find matches. It is a test of declarative memory. The one with the most correct pairs wins the game.

In the experiment, the control groups played the game normally. But the experimental groups didn't. They played the game in the presence of rose scent. Then everybody went to bed. The control groups were allowed to sleep unperturbed. Soon after the snoring began in the experimental groups, however, the researchers filled their rooms with the same rose scent. Upon awakening, the subjects were tested on their knowledge of where the matches had been discovered on the previous day. Those subjects without the scent answered correctly 86 percent of the time. Those re-exposed to the scent answered correctly 97 percent of the time. Brain imaging experiments showed the direct involvement of the hippocampus. It is quite possible that the smell enhanced recall during the offline processing that normally occurs during sleep.

In the highly competitive world of school performance, there are parents who would die to give their kids an 11 percent edge over the competition. Some CEOs would appreciate such an advantage, too, in the face of anxious shareholders.

Sensory branding

Author Judith Viorst once said, "Strength is the capacity to break a chocolate bar into four pieces, and then eat just one of the pieces." She was of course referring to the power of the confection on self-will. It's a testament to the power of emotion to incite action.

That's just what emotions do: affect motivations. As we discussed in the Attention chapter, emotions are used by the brain to select certain inputs for closer inspection. Because smells stimulate areas in the brain responsible for creating emotions as well as memories, a number of business people have asked, "Can smell, which can affect motivation, also affect sales?"

One company tested the effects of smell on business and found a whopper of a result. Emitting the scent of chocolate from a vending machine, it found, drove chocolate sales up 60 percent. That's quite a motivation. The same company installed a waffle-cone-smell emitter near a location-challenged ice cream shop (it was inside a large hotel and hard to find). Sales soared 50 percent, leading the inventor to coin the term "aroma billboard" to describe the technique.

Welcome to the world of sensory branding. An entire industry is beginning to pay attention to human sensory responses, with smell as the centerpiece. In an experiment for a clothing store, investigators subtly wafted the smell of vanilla in the women's department, a scent known to produce a positive response among women. In the men's department, they diffused the smell of rose maroc, a spicy, honeylike fragrance that had been pretested on men. The retail results were amazing. When the scents were deployed, sales doubled from their typical average in each department. And when the scents were reversed—vanilla for men and rose maroc for women—sales plummeted below their typical average. The conclusion? Smell works, but only when deployed in a particular way. "You can't just use a pleasant scent and expect it to work," says Eric Spangenberg, the scientist in charge of the work. "It has to be congruent." In recognition of this fact, Starbucks does not allow employees to wear perfume on company time. It interferes with the seductive smell of the coffee they serve and its potential to attract customers.

Marketing professionals have begun to come up with recommendations for the use of smell in differentiating a brand: First, match the scent with the hopes and needs of the target market. The pleasant smell of coffee may remind a busy executive of the comforts of home, a welcome relief when about to close a deal. Second, integrate the odor with the "personality" of the object for sale. The fresh smell of a forest, or the salty odor of a beach, might evoke a sense of adventure more so than, say, the smell of vanilla, in potential buyers of SUVs. Remember the Proust effect, that smell can evoke memory.

Smells at work (not coming from the fridge)

What about the role of learning in a business setting? Two ideas come to mind, based loosely on my teaching experiences. I occasionally teach a molecular biology class for engineers, and one time I decided to do my own little Proust experiment. (There was nothing rigorous about this little parlor investigation; it was simply an informal inquiry.) Every time I taught one section on an enzyme (called RNA polymerase II), I prepped the room by squirting the perfume Brut on one wall. In an identical class in another building, I taught the same material, but I did not squirt Brut when describing the enzyme. Then I tested everybody, squirting the perfume into both classrooms. Every time I did this experiment, I got the same result. The people who were exposed to the perfume during learning did better on subject matter pertaining to the enzyme—sometimes dramatically better—than those who were not.

And that led me to an idea. Many businesses have a need to teach their clients about their products, from how to implement software to how to repair airplane engines. For financial reasons, the classes are often compressed for time and packed with information, 90 percent of which is forgotten a *day* later. (For most declarative subjects, memory degradation starts the first few hours after the teaching is finished.) But what if the teacher paired a smell with each lesson, as in my Brut experiment? One might even expose the students to the smell while they are asleep. The students could not help but associate the autobiographical experience of the class—complete with the intense transfer of information—with the odorant.

After the class, the students (let's say they're learning to repair airplane engines) return to their company. Two weeks later, they are confronted with a room full of newly broken engines to repair. Most of them will have forgotten something in the intense class they took and need to review their notes. This review would take place in the presence of the smell they encountered during the learning. Would it give a boost to their memories? What if they were exposed to the smell while they were in the shop repairing the actual engines? The enhanced memory might improve performance, even cut down on errors.

Sound preposterous? Possibly. Indeed, one must be careful to tease out context-dependent learning (remember those dive suits from Chapter 5) from true multisensory environments. But it's a start toward thinking about learning environments that go beyond the normal near-addiction to visual and auditory information. It is an area where much potential research fruit lies—truly a place for brain scientists, educators and business professionals to work together in a practical way.

Summary

Rule #9
Stimulate more of the senses at the same time.

* We absorb information about an event through our senses, translate it into electrical signals (some for sight, others from sound, etc.), disperse those signals to separate parts of the brain, then reconstruct what happened, eventually perceiving the event as a whole.

* The brain seems to rely partly on past experience in deciding how to combine these signals, so two people can perceive the same event very differently.

* Our senses evolved to work together—vision influencing hearing, for example—which means that we learn best if we stimulate several senses at once.

* Smells have an unusual power to bring back memories, maybe because smell signals bypass the thalamus and head straight to their destinations, which include that supervisor of emotions known as the amygdala.

Get more at www.brainrules.net

vision

Rule #10
Vision trumps all other senses.

WE DO NOT SEE with our eyes. We see with our brains.

The evidence lies with a group of 54 wine aficionados. Stay with me here. To the untrained ear, the vocabularies that wine tasters use to describe wine may seem pretentious, more reminiscent of a psychologist describing a patient. ("Aggressive complexity, with just a subtle hint of shyness" is something I once heard at a wine-tasting soirée to which I was mistakenly invited—and from which, once picked off the floor rolling with laughter, I was hurriedly escorted out the door).

These words are taken very seriously by the professionals, however. A specific vocabulary exists for white wines and a specific vocabulary for red wines, and the two are never supposed to cross. Given how individually we each perceive any sense, I have often wondered how objective these tasters actually could be. So, apparently, did a group of brain researchers in Europe. They descended upon ground zero of the wine-tasting world, the University of Bordeaux, and asked: "What if we dropped odorless, tasteless red dye into white wines, then gave it to 54 wine-tasting professionals?" With only visual sense altered, how would the enologists now describe their wine? Would their delicate palates see through the ruse, or would their noses be fooled? The answer is "their noses would be fooled." When the wine tasters encountered the altered whites, every one of them employed the vocabulary of the reds. The visual inputs seemed to trump their other highly trained senses.

Folks in the scientific community had a field day. Professional research papers were published with titles like "The Color of Odors" and "The Nose Smells What the Eye Sees." That's about as much frat boy behavior as prestigious brain journals tolerate, and you can almost see the wicked gleam in the researchers' eyes. Data such as these point to the nuts and bolts of this chapter's Brain Rule. Visual processing doesn't just assist in the perception of our world. It dominates the perception of our world. Starting with basic biology, let's find out why.

a hollywood horde

We see with our brains.

This central finding, after years of study, is deceptively simple. It is made more misleading because the internal mechanics of vision seem easy to understand. First, light (groups of photons, actually) enters our eyes, where it is bent by the cornea, the fluid-filled structure upon which your contacts normally sit. Then the light travels through the eye to the lens, where it is focused and allowed to strike the retina, a group of neurons in the back of the eye. The collision generates electric signals in these cells, and the signals travel deep into the brain via the optic nerve. The brain then interprets the electrical information, and we become visually aware. These steps seem effortless, 100 percent trustworthy, capable of providing a completely accurate representation of what's actually out there.

Though we are used to thinking about our vision in such reliable terms, nothing in that last sentence is true. The process is extremely complex, seldom provides a completely accurate representation of our world, and is not 100 percent trustworthy. Many people think that the brain's visual system works like a camera, simply collecting and processing the raw visual data provided by our outside world. Such analogies mostly describe the function of the eye, however, and not particularly well. We actually experience our visual environment as a fully analyzed *opinion* about what the brain thinks is out there.

We thought that the brain processed information such as color, texture, motion, depth, and form in discrete areas; higher-level structures in the brain then gave meaning to these features, and we suddenly obtained a visual perception. This is very similar to the steps discussed in the Multisensory chapter: sensing, routing, and perception, using bottom-up and top-down methods. It is becoming clearer that we need to amend this notion. We now know that visual analysis starts surprisingly early on, beginning when light strikes the retina. In the old days, we thought this collision was a mechanical, automated process: A photon shocked a retinal nerve cell into cracking off some electrical signal, which eventually found its way to the back of our heads. All perceptual heavy lifting was done afterward, deep in the bowels of the brain. There is strong evidence that this is not only a simplistic explanation of what goes on. It is a wrong explanation.

Rather than acting like a passive antenna, the retina appears to quickly process the electrical patterns before it sends anything off to Mission Control. Specialized nerve cells deep within the retina interpret the patterns of photons striking the retina, assemble the patterns into partial "movies," and then send these movies off to the back of our heads. The retina, it seems, is filled with teams of tiny Martin Scorseses. These movies are called tracks. Tracks are coherent, though partial, abstractions of specific features of the visual environment. One track appears to transmit a movie you might call *Eye Meets Wireframe*. It is composed only of outlines, or edges. Another makes a film you might call *Eye Meets Motion*, processing only the movement of an object (and often in a specific direction). Another makes *Eye Meets Shadows*. There may be as many as 12 of these tracks operating simultaneously in the retina, sending off interpretations of specific features of the visual field. This new view is quite unexpected. It's like discovering that the reason your TV gives you feature films is that your cable is infested by a dozen amateur independent filmmakers, hard at work creating the feature while you watch it.

streams of consciousness

These movies now stream out from the optic nerve, one from each eye, and flood the thalamus, that egg-shaped structure in the middle of our heads that serves as a central distribution center for most of our senses. If these streams of visual information can be likened to a large, flowing river, the thalamus can be likened to the beginning of a delta. Once it leaves the thalamus, the information travels along increasingly divided neural streams. Eventually, there will be thousands of small neural tributaries carrying parts of the original information to the back of the brain. The information drains into a large complex region within the occipital lobe called the visual cortex. Put your hand on the back of your head. Your palm is now less than a quarter of an inch away from the area of the brain that is currently allowing you to see this page. It is a quarter of an inch away from your visual cortex.

The visual cortex is a big piece of neural acreage, and the various streams flow into specific parcels. There are thousands of lots, and their functions are almost ridiculously specific. Some parcels respond only to diagonal lines, and only to specific diagonal lines (one region responds to a line tilted at 40 degrees, but not to one tilted at 45). Some process only the color information in a visual signal; others, only edges; others, only motion.

Damage to the region responding to motion results in an extraordinary deficit: the inability to see moving objects as actually moving. This can be very dangerous, observable in the famous case of a Swiss woman we'll call Gerte. In most respects, Gerte's eyesight was normal. She could provide the names of objects in her visual field; recognize people, both familiar and unfamiliar, as human; read newspapers with ease. But if she looked at a horse galloping across a field, or a truck roaring down the freeway, she saw no motion. Instead, she saw a sequence of static, strobe-like snapshots of the objects. There was no smooth impression of continuous motion, no effortless perception of instantaneous changes of location. There was no motion of any kind. Gerte became terrified to cross the street. Her strobe-like world did not allow her to calculate the speed or destination of the vehicles. She could not perceive the cars as moving, let alone moving toward her (though she could readily identify the offending objects as automobiles, down to make and license plate). Gerte even said that talking to someone face-to-face was like speaking on the phone. She could not see the changing facial expressions associated with normal conversation. She could not see "changing" at all.

Gerte's experience shows the modularity of visual processing. But it is not just motion. Thousands of streams feeding into these regions allow for the separate processing of individual features. And if that was the end of the visual story, we might perceive our world with the unorganized fury of a Picasso painting, a nightmare of fragmented objects, untethered colors, and strange, unboundaried edges.

But that's not what happens, because of what takes place next. At the point where the visual field lies in its most fragmented state, the brain decides to reassemble the scattered information. Individual tributaries start recombining, merging, pooling their information, comparing their findings, and then sending their analysis to higher brain centers. The centers gather these hopelessly intricate calculations from many sources and integrate them at an even more sophisticated level. Higher and higher they go, eventually collapsing into two giant streams of processed information. One of these, called the ventral stream, recognizes what an object is and what color it possesses. The other, termed the dorsal stream, recognizes the location of the object in the visual field and whether it is moving. "Association regions" do the work of integrating the signals. They associate—or, better to say, reassociate—the balkanized electrical signals. Then, you see something. So, the process of vision is not as simple as a camera taking a picture. The process is more complex and more convoluted than anyone could have imagined. There is no real scientific agreement about why this disassembly and reassembly strategy even occurs.

Complex as visual processing is, things are about to get worse. We generally trust our visual apparati to serve us a faithful, up-to-the-minute, 100 percent accurate representation of what's actually out there. Why do we believe that? Because our brain insists on helping us create our perceived reality. Two examples explain this exasperating tendency. One involves people who see miniature policemen who aren't there. The other involves the active perception of camels.

camels and cops

You might inquire whether I had too much to drink if I told you right now that you were actively hallucinating. But it's true. At this very moment, while reading this text, you are perceiving parts of this page that do not exist. Which means you, my friend, are hallucinating. I am about to show you that your brain actually likes to make things up, not 100 percent faithful to what the eyes broadcast to it.

There is a region in the eye where retinal neurons, carrying visual information, gather together to begin their journey into deep brain tissue. That gathering place is called the optic disk. It's a strange region, because there are no cells that can perceive sight in the optic disk. It is blind in that region—and you are, too. It is called the blind spot, and each eye has one. Do you ever see two black holes in your field of view that won't go away? That's what you should see. But your brain plays a trick on you. As the signals are sent to your visual cortex, the brain detects the presence of the holes and then does an extraordinary thing. It examines the visual information 360 degrees around the spot and calculates what is most likely to be there. Then, like a paint program on a computer, it fills in the spot. The process is called "filling in," but it could be called "faking it." Some believe that the brain simply ignores the lack of visual information, rather than calculating what's missing. Either way, you're not getting a 100 percent accurate representation.

It should not be surprising that the brain possesses such independent-minded imaging systems. Proof is as close as last night's dream. But just how much of a loose cannon these systems can be is evidenced in a phenomenon known as the Charles Bonnet Syndrome. Millions of people suffer from it. Most who have it keep their mouth shut, however, and perhaps with good reason. People with Charles Bonnet syndrome see things that aren't there. It's like the blind-spot-fill-in apparatus gone horribly wrong. For some patients with Charles Bonnet, everyday household objects suddenly pop into view. For others, unfamiliar people unexpectedly appear next to them at dinner. Neurologist Vilayanur Ramachandran describes the case of a woman who suddenly—and delightfully—observed two tiny policemen scurrying across the floor, guiding an even smaller criminal to a matchbox-size van. Other patients have reported angels, goats in overcoats, clowns, Roman chariots, and elves. The illusions often occur in the evening and are usually quite benign. It is common among the elderly, especially among those who previously suffered damage somewhere in their visual pathway. Extraordinarily, almost all of the patients experiencing the hallucinations know that they aren't real. No one really knows why they occur.

This is just one example of the powerful ways brains participate in our visual experience. Far from being a camera, the brain is actively deconstructing the information given to it by the eyes, pushing it through a series of filters, and then reconstructing what it thinks it sees. Or what it thinks you should see.

Yet even this is hardly the end of the mystery. Not only do you perceive things that aren't there with careless abandon, but exactly *how* you construct your false information follows certain rules. Previous experience plays an important role in what the brain allows you to see, and the brain's assumptions play a vital role in our visual perceptions. We consider these ideas next.

Since ancient times, people have wondered why two eyes give rise to a single visual perception. If there is a camel in your left eye and a camel in your right eye, why don't you perceive two camels? Here's an experiment to try that illustrates the problem nicely.

1) Close your left eye, then stretch your left arm in front of you.

2) Raise up the index finger of your left hand, as if you were pointing to the sky.

3) Keep the arm in this position while you hold your right arm about six inches in front of your face. Raise your right index finger like it too was pointing to the sky.

4) With your eye still closed, position your right index finger so that it appears just to the left of your left index finger.

5) Now speedily open you left eye and close the right one. Do this several times.

If you positioned your fingers correctly, your right finger will jump to the other side of your left finger and back again. When you open both eyes, the jumping will stop. This little experiment shows that the two images appearing on each retina always differ. It also shows that both eyes working together somehow give the brain enough information to see non-jumping reality.

Why do you see only one camel? Why do you see two arms

with stable, non-jumping fingers? Because the brain interpolates the information coming from both eyes. It makes about a gazillion calculations, then provides you its best guess. And it is a guess. You can actually show that the brain doesn't really know where things are. Rather, it hypothesizes the probability of what the current event should look like and then, taking a leap of faith, approximates a viewable image. What you experience is not the image. What you experience is the leap of faith. Why does the brain do this? Because it is forced to solve a problem: We live in a three-dimensional world, but the light falls on our retina in a two-dimensional fashion. The brain must deal with this disparity if it is going to accurately portray the world. Just to complicate things, our two eyes give the brain two separate visual fields, and they project their images upside down and backward. To make sense of it all, the brain is forced to start guessing.

Upon what does it base its guesses, at least in part? The answer is bone-chilling: prior experience with events in your past. After adamantly inserting numerous assumptions about the received information (some of these assumptions may be inborn), the brain then offers up its findings for your perusal. It goes to all of this trouble for an important reason dripping with Darwinian good will: so you will see one camel in the room when there really is only one camel in the room (and see its proper depth and shape and size and even hints about whether or not it will bite you). All of this happens in about the time it takes to blink your eyes. Indeed, it is happening right now.

If you think the brain has to devote to vision a lot of its precious thinking resources, you are right on the money. It takes up about half of everything you do, in fact. This helps explains why snooty wine tasters with tons of professional experience throw out their taste buds so quickly in the thrall of visual stimuli. And that lies at the very heart of this chapter's Brain Rule.

phantom of the ocular

In the land of sensory kingdoms, there are many ways to show that vision isn't the benevolent prime minister but the dictatorial emperor. Take phantom-limb experiences. Sometimes, people who have suffered an amputation continue to experience the presence of their limb, even though no limb exists. Sometimes the limb is perceived as frozen into a fixed position. Sometimes it feels pain. Scientists have used phantoms to demonstrate the powerful influence vision has on our senses.

An amputee with a "frozen" phantom arm was seated at a table upon which had been placed a topless, divided box. There were two portals in the front, one for the arm and one for the stump. The divider was a mirror, and the amputee could view a reflection of either his functioning hand or his stump. When he looked at his functioning hand, he could see his right arm present and his left arm missing. But when he looked at the reflection of his right arm in the mirror—what looked like another arm—the phantom limb on the other side of the box suddenly "woke up." If he moved his normal hand while gazing at its reflection, he could feel his phantom move, too. And when he stopped moving his right arm, his missing left arm "stopped" also. The addition of visual information began convincing his brain of a miraculous rebirth of the absent limb. This is vision not only as dictator but as faith healer. The visual-capture effect is so powerful, it can be used to alleviate pain in the phantom.

How do we measure vision's dominance?

One way is to show its effects on learning and memory. Researchers historically have used two types of memory in their investigations. The first, recognition memory, is a glorified way to explain familiarity. We often deploy recognition memory when looking at old family photographs, such as gazing at a picture of an old aunt not remembered for years. You don't necessarily recall her name, or the photo, but you still recognize her as your aunt. You may not be able to recall certain details, but as soon as you see it, you know that you have seen it before.

Other types of learning involve the familiar working memory. Explained in greater detail in the Memory chapters, working memory is that collection of temporary storage buffers with fixed capacities and frustratingly short life spans. Visual short-term memory is the slice of that buffer dedicated to storing visual information. Most of us can hold about four objects at a time in that buffer, so it's a pretty small space. And it appears to be getting smaller. Recent data show that as the complexity of the objects increases, the number of objects capable of being captured drops. The evidence also suggests that the number of objects and complexity of objects are engaged by different systems in the brain, turning the whole notion of short-term capacity, if you will forgive me, on its head. These limitations make it all the more remarkable—or depressing—that vision is probably the best single tool we have for learning anything.

worth a thousand words

When it comes to memory, researchers have known for more than 100 years that pictures and text follow very different rules. Put simply, the more visual the input becomes, the more likely it is to be recognized—and recalled. The phenomenon is so pervasive, it has been given its own name: the pictorial superiority effect, or PSE.

Human PSE is truly Olympian. Tests performed years ago showed that people could remember more than 2,500 pictures with at least 90 percent accuracy several days post-exposure, even though subjects saw each picture for about 10 seconds. Accuracy rates a year later still hovered around 63 percent. In one paper—adorably titled "Remember Dick and Jane?"—picture recognition information was reliably retrieved several decades later.

Sprinkled throughout these experiments were comparisons with other forms of communication. The favorite target was usually text or oral presentations, and the usual result was "picture demolishes them both." It still does. Text and oral presentations are not just less efficient than pictures for retaining certain types of information; they are *way* less efficient. If information is presented orally, people remember about 10 percent, tested 72 hours after exposure. That figure goes up to 65 percent if you add a picture.

The inefficiency of text has received particular attention. One of the reasons that text is less capable than pictures is that the brain sees words as lots of tiny pictures. Data clearly show that a word is unreadable unless the brain can separately identify simple features in the letters. Instead of words, we see complex little art-museum masterpieces, with hundreds of features embedded in hundreds of letters. Like an art junkie, we linger at each feature, rigorously and independently verifying it before moving to the next. The finding has broad implications for reading efficiency. Reading creates a bottleneck. My text chokes you, not because my text is not enough like pictures but because my text is too much like pictures. To our cortex, unnervingly, there is no such thing as words.

That's not necessarily obvious. After all, the brain is as adaptive as Silly Putty. With years of reading books, writing email, and sending text messages, you might think the visual system could be trained to recognize common words without slogging through tedious additional steps of letter-feature recognition. But that is not what happens. No matter how experienced a reader you become, you will still stop and ponder individual textual features as you plow through these pages, and you will do so until you can't read anymore.

Perhaps, with hindsight, we could have predicted such inefficiency. Our evolutionary history was never dominated by text-filled billboards or Microsoft Word. It was dominated by leaf-filled trees and saber-toothed tigers. The reason vision means so much to us may be as simple as the fact that most of the major threats to our lives in the savannah were apprehended visually. Ditto with most of our food supplies. Ditto with our perceptions of reproductive opportunity.

The tendency is so pervasive that, even when we read, most of us try to visualize what the text is telling us. "Words are only postage stamps delivering the object for you to unwrap," George Bernard Shaw was fond of saying. These days, there is a lot of brain science technology to back him up.

a punch in the nose

Here's a dirty trick you can pull on a baby. It may illustrate something about your personality. It certainly illustrates something about visual processing.

Tie a ribbon around the baby's leg. Tie the other end to a bell. At first she seems to be randomly moving her limbs. Soon, however, the infant learns that if she moves one leg, the bell rings. Soon she is happily—and preferentially—moving that leg. The bell rings and rings and rings. Now cut the ribbon. The bell no longer rings. Does that stop the baby? No. She still kicks her leg. Something is wrong, so she kicks harder. Still no sound. She does a series of rapid kicks in sequence. Still no success. She gazes up at the bell, even stares at the bell. This visual behavior tells us she is paying attention to the problem. Scientists can measure the brain's attentional state even with the diaper-and-breast-milk crowd because of this reliance on visual processing.

This story illustrates something fundamental about how brains perceive their world. As babies begin to understand cause-and-effect relationships, we can determine how they pay attention by watching them stare at their world. The importance of this gazing behavior cannot be underestimated. Babies use visual cues to show they are paying attention to something—even though nobody taught them to do that. The conclusion is that babies come with a variety of preloaded software devoted to visual processing.

That turns out to be true. Babies display a preference for patterns with high contrast. They seem to understand the principle of common fate: Objects that move together are perceived as part of the same object, such as stripes on a zebra. They can discriminate human faces from non-human equivalents and seem to prefer them. They possess an understanding of size related to distance—that if an object is getting closer (and therefore getting bigger), it is still the same object. Babies can even categorize visual objects by common physical characteristics. The dominance that vision displays behaviorally begins in the tiny world of infants.

And it shows up in the even tinier world of DNA. Our sense of smell and color vision are fighting violently for evolutionary control, for the right to be consulted first whenever something on the outside happens. And vision is winning. In fact, about 60 percent of our smell-related genes have been permanently damaged in this neural arbitrage, and they are marching toward obsolescence at a rate fourfold faster than any other species sampled. The reason for this decommissioning is simple: The visual cortex and the olfactory cortex take up a lot of neural real estate. In the crowded zero-sum world of the sub-scalp, something has to give.

Whether looking at behavior, cells, or genes, we can observe how important visual sense is to the human experience. Striding across our brain like an out-of-control superpower, giant swaths of biological resource are consumed by it. In return, our visual system creates movies, generates hallucinations, and consults with previous information before allowing us to see the outside. It happily bends the information from other senses to do its bidding and, at least in the case of smell, seems to be caught in the act of taking it over.

Is there any point in trying to ignore this juggernaut, especially if you are a parent, educator, or business professional? You don't have to go any further than the wine experts of Bordeaux for proof.

ideas

I owe my career choice to Donald Duck. I am not joking. I even remember the moment he convinced me. I was 8 years old at the time, and my mother trundled the family off to a showing of an amazing 27-minute animated short called *Donald in Mathmagic Land*. Using visual imagery, a wicked sense of humor, and the wide-eyed wonder of an infant, Donald Duck introduced me to math. Got me excited about it. From geometry to football to playing billiards, the power and beauty of mathematics were made so real for this nerd-in-training, I asked if I could see it a second time. My mother obliged, and the effect was so memorable, it eventually influenced my career choice. I now have a copy of those valuable 27 minutes in my own home and regularly inflict it upon my poor children. *Donald in Mathmagic Land* won an Academy Award for best animated short of 1959. It also should have gotten a "Teacher of the Year" award. The film illustrates—literally—the power of the moving image in communicating complex information to students. It's one inspiration for these suggestions.

Teachers should learn why pictures grab attention

Educators should know how pictures transfer information. There are things we know about how pictures grab attention that are rock solid. We pay lots of attention to color. We pay lots of attention to orientation. We pay lots of attention to size. And we pay special attention if the object is in motion. Indeed, most of the things that threatened us in the Serengeti *moved*, and the brain has evolved unbelievably sophisticated trip-wires to detect it. We even have specialized regions to distinguish when our eyes are moving versus when our world is moving. These regions routinely shut down perceptions of eye movement in favor of the environmental movement.

Teachers should use computer animations

Animation captures the importance not only of color and placement but also of motion. With the advent of web-based graphics, the days when this knowledge was optional for educators are probably over. Fortunately, the basics are not hard to learn. With today's software, simple animations can be created by anybody who knows how to draw a square and a circle. Simple, two-dimensional pictures are quite adequate; studies show that if the drawings are too complex or lifelike, they can distract from the transfer of information.

Test the power of images

Though the pictorial superiority effect is a well-established fact for certain types of classroom material, it is not well-established for all material. Data are sparse. Some media are better at communicating some types of information than others. Do pictures communicate conceptual ideas such as "freedom" and "amount" better than, say a narrative? Are language arts better represented in picture form, or are other media styles more robust? Working out these issues in real-world classrooms would provide the answer, and that takes collaboration between teachers and researchers.

Communicate with pictures more than words

"Less text, more pictures" were almost fighting words in 1982. They were used derisively to greet the arrival of USA Today, a brand-new type of newspaper with, as you know, less text, more pictures. Some predicted the style would never work. Others predicted that if it did, the style would spell the end of Western civilization as the newspaper-reading public knows it. The jury may be out on the latter prediction, but the former has a powerful and embarrassing verdict. Within four years, USA Today had the second highest readership of any newspaper in the country, and within 10, it was the number one. It still is.

What happened? First, we know that pictures are a more efficient delivery mechanism of information than text. Second, the American work force is consistently overworked, with more things being done by fewer people. Third, many Americans still read newspapers. In the helter-skelter world of overworked Americans, more-efficient information transfer may be the preferred medium. As the success of *USA Today* suggests, the attraction may be strong enough to persuade consumers to reach for their wallet. So, pictorial information may be initially more attractive to consumers, in part because it takes less effort to comprehend. Because it is also a more efficient way to glue information to a neuron, there may be strong reasons for entire marketing departments to think seriously about making pictorial presentations their primary way of transferring information.

The initial effect of pictures on attention has been tested. Using infrared eye-tracking technology, 3,600 consumers were tested on 1,363 print advertisements. The conclusion? Pictorial information was superior in capturing attention—independent of its size. Even if the picture was small and crowded with lots of other non-pictorial elements close to it, the eye went to the visual. The researchers in the study, unfortunately, did not check for retention.

Toss your PowerPoint presentations

The presentation software called PowerPoint has become ubiquitous, from corporate boardrooms to college classrooms to scientific conferences. What's wrong with that? It's text-based, with six hierarchical levels of chapters and subheads—all words. Professionals everywhere need to know about the incredible inefficiency of text-based information and the incredible effects of images. Then they need to do two things:

1) Burn their current PowerPoint presentations.

2) Make new ones.

Actually, the old ones should be stored, at least temporarily, as useful comparisons. Business professionals should test their new designs against the old and determine which ones work better. A typical PowerPoint business presentation has nearly 40 words *per slide*. That means we have a lot of work ahead of us.

Summary

Rule #10
Vision trumps all other senses.

• Vision is by far our most dominant sense, taking up half of our brain's resources.

• What we see is only what our brain tells us we see, and it's not 100 percent accurate.

• The visual analysis we do has many steps. The retina assembles photons into little movie-like streams of information. The visual cortex processes these streams, some areas registering motion, others registering color, etc. Finally, we combine that information back together so we can see.

• We learn and remember best through pictures, not through written or spoken words.

Get more at www.brainrules.net

gender

Rule #11
Male and female brains are different.

THE MAN WAS A hot dog. The woman was a bitch.

The results of the experiment could be summarized in those two sentences. Three researchers created a fictitious assistant vice president of an aircraft company. Four groups of experimental subjects, with equal numbers of men and women in each group, were asked to rate this fictional person's job performance. Each group was given the vice president's brief job description, but the first group also was told that the vice president was a man. They were asked to rate both the competence and the likability of the candidate. They gave a very flattering review, rating the man "very competent" and "likable." The second group was told that the vice president was a woman. She was rated "likable" but "not very competent." All other factors were equal. Only the perceived gender had changed.

The third group was told that the vice president was a male superstar, a stellar performer on the fast track at the company. The fourth group was told that the vice president was a female superstar, also on the express lane to the executive washroom. As before, the third group rated the man "very competent" and "likable." The woman superstar also was rated "very competent." But she was not rated "likable." In fact, the group's descriptions included words such as "hostile." As I said, the man was a hot dog. The woman was a bitch.

The point is, gender biases hurt real people in real-world situations. As we hurtle headlong into the controversial world of brains and genders, keeping these social effects in mind is excruciatingly important. There is a great deal of confusion regarding the way men and women relate to each other, and even more about why. There is confusion about the terms as well, blurring the line between the concepts of "sex" and "gender." Here, sex will generally refer to biology and anatomy. Gender will refer mostly to social expectations. Sex is set into the concrete of DNA. Gender is not. The differences between men's and women's brains start with how they got that way in the first place.

the x factor

How do we become male and female? The road to sex assignment starts out with all the enthusiasm sex usually stimulates. Four hundred million sperm fall all over themselves attempting to find one egg during intercourse. The task is not all that difficult. In the microscopic world of human fertilization, the egg is the size of the Death Star, and the sperm are the size of X-wing fighters. X is a good letter for this enterprise—the name of that very important chromosome that half of all sperm and all eggs carry. You recall chromosomes from biology class, those writhing strings of DNA packed into the nucleus that contain the information necessary to make you. It takes 46 of them to do it, which you can think of as 46 volumes in an encyclopedia. Twenty-three come from Mom, and 23 come from Dad. Two are sex chromosomes. At least one of those chromosomes has to be an X chromosome, or you will die.

If you get two X chromosomes, you go into the ladies locker room all your life; an X and Y puts you forever in the men's. This sex assignment is controlled by the male. Henry VIII's wives wish he'd known that. He executed one of them for being unable to produce a boy as heir to the throne, but he should have executed himself. The Y can be donated only by sperm (the egg never carries one), so the male determines the sex.

Gender differences can be divided into three areas: genetic, neuroanatomical, and behavioral. Scientists usually spend their whole careers exploring only one—each difference is like a separate island in a common research ocean. We'll tour all three, starting with a molecular explanation of why Henry VIII owes Anne Boleyn a big fat apology.

One of the most interesting facts about the Y chromosome is that you don't need most of it to make a male. All it takes to kick-start the male developmental program is a small snippet near the middle, carrying a gene called SRY. In our tour, we immediately notice Gene Island is dominated by a single scientist, David C. Page. He is the researcher who isolated SRY. Though in his 50s, Page looks to be about 28 years old. As director of the Whitehead Institute and a professor at MIT, he is a man of considerable intellect. He also is charming and has a refreshingly wicked sense of humor. Page is the world's first molecular sex therapist. Or, better, sex broker. He discovered that you can destroy the SRY gene in a male embryo and get a female, or add SRY to a female embryo and turn her into a male (SR stands for "sex reversal"). Why can you do this? In a fact troubling to anybody who believes males are biologically hard-wired to dominate the planet, researchers discovered that the basic default setting of the mammalian embryo is to become female.

There is terrible inequality between the two chromosomes. The X chromosome does most of the heavy developmental lifting, while the little Y has been shedding its associated genes at a rate of about five every one million years, committing suicide in slow motion. It's now down to less than 100 genes. By comparison, the X chromosome carries about 1,500 genes, all necessary participants in embryonic construction projects. These are not showing any signs of decay.

With only a single X chromosome, males need every X gene they can get. Females, however, have double the necessary amount. You can think of it like a cake recipe calling for only one cup of flour. If you decide to put in two, things will change in a most unpleasant fashion. The female embryo uses what may be the most time-honored weapon in the battle of the sexes to solve the problem of two X's: She simply ignores one of them. This chromosomal silent treatment is known as X inactivation. One of the chromosomes is tagged with the molecular equivalent of a "Do Not Disturb" sign. Since there are two X's from which to choose, Mom's or Dad's, researchers wanted to know who preferentially got the sign.

The answer was completely unexpected. *There were no preferences.* Some cells in the developing little girl embryo hung their sign around Mom's X. Neighboring cells hung their sign around Dad's. At this point in research, there doesn't appear to be any rhyme or reason, and it is considered a random event. This means that cells in the female embryo are a complex mosaic of both active and inactive mom-and-pop X genes. Because males require all 1,500 X genes to survive, and they have only one X-chromosome, it would be stupid for them to hang up "Do Not Disturb" notes. They never do it. X inactivation does not occur in guys. And because males must get their X from Mom, all men are literally, with respect to their X chromosome, Momma's Boys—unisexed. That's very different from their sisters, who are more genetically complex. These bombshells describe our first truly genetic-based findings of potential gender differences.

We now know the function of many of the 1,500 genes that reside on the X chromosome. Swallow hard here. Many of those genes involve brain function. Many of them govern how we *think*. In 2005, the human genome was sequenced, and an unusually large percentage of the X chromosome genes were found to create proteins involved in brain manufacture. Some of these genes may be involved in establishing higher cognitive functions, from verbal skills and social behavior to certain types of intelligence. Researchers call the X chromosome a cognitive "hot spot."

These findings represent one of the most important regions on Gene Island. But it is hardly the only important region, and not even the most important island.

is bigger better?

The purpose of genes is to create molecules that mediate the functions of the cells in which they reside. Collections of these cells create the neuroanatomy of the brain (which in turn creates our behavior). Leaving Gene Island, our next stop is Cell Island, a region where scientists investigate large structures in the brain, or neuroanatomy. Here, the real trick is finding structures that *aren't* affected by sex chromosome dosage.

Labs—headed by scientists of both sexes, I should perhaps point out—have found differences in the front and prefrontal cortex, areas of the brain that control much of our decision-making ability. This cortex is fatter, in certain parts, in women than in men. There are sex-based differences in the limbic system, which controls our emotional life and mediates some types of learning. Prominent differences lie in the amygdala, controlling not only the generation of emotions but also the ability to remember them. Running counter to current social prejudice, this region is much larger in men than it is in women. At rest, female amygdalas tend to talk mostly to the left hemisphere, while male amygdalas do most of their chatting with the right hemisphere. Brain cells communicate via biochemicals, and these have not escaped sex differences, either. The regulation of serotonin is particularly dramatic. Serotonin is key in regulating emotion and mood (Prozac works by altering the regulation of this neurotransmitter). Males can synthesize serotonin about 52 percent faster than females. Do these physical differences mean anything? In animals, the size of structures is thought to reflect their relative importance to survival. Human examples at first blush seem to follow a similar pattern. We already have noticed that violinists have bigger areas of the brain devoted to controlling their left hand than their right. But neuroscientists nearly come to blows over how structure relates to function. We don't yet know whether differences in neurotransmitter distributions, or in the size of a brain region, mean anything substantial.

Such cautions have not stopped brain scientists from going after the question of behavior differences, and they won't stop us, either. Fasten your seat belts and strap on the Kevlar, for we are about to land on the noisiest, most intellectually violent island on our imaginary itinerary: Behavior Island.

battle of the sexes

I didn't really want to write about this. Characterizing gender-specific behaviors has a long and mostly troubled history. Even institutions holding our best minds aren't immune. Larry Summers was *Harvard's* president, for Pete's sake, when he attributed girls' lower math and science scores to behavioral genetics, comments that cost him his job. And he is in exceptionally good intellectual company. Consider these three quotes:

"The female is an impotent male, incapable of making semen because of the coldness of her nature. We therefore should look upon the female state as if it were a deformity, though one that occurs in the ordinary course of nature."

Aristotle (384–332 BC)

"Girls begin to talk and to stand on their feet sooner than boys because weeds always grow up more quickly than good crops."

Martin Luther (1483–1546)

"If they can put a man on the moon ... why can't they put them all there?"
Jill (1985, graffiti on a bathroom wall,
in response to Luther's quote)

And so the weary battle of the sexes continues. Almost 2,400 years of history separate Aristotle from Jill, yet we seem to have barely moved. Invoking planet metaphors like Venus and Mars, some purport to expand perceived differences into prescriptions for relationships. And this is the most scientifically progressive era in human history.

Mostly, I think, it comes down to statistics.

There may very well be differences in the way men and women think about some things. But when people hear about measurable differences, they often think scientists are talking about individuals, such as themselves. That's a *big* mistake. When scientists look for behavioral trends, they do not look at individuals. They look at populations. Statistics in these studies can never apply to individuals. Trends emerge, but there are variations within a population, often with significant overlaps between the genders. It is true that every time neuroscientist Flo Haseltine does an fMRI, she sees different parts of the brain light up depending upon whether she is viewing a man or a woman. Exactly how that relates to your behavior is a completely separate question.

first hints

What we do know about the biological roots of behavioral differences began with brain pathologies. Mental retardation is more common in males than in females in the general population. Many of these pathologies are caused by mutations in any one of 24 genes within the X chromosome. As you know, males have no backup X. If their X gets damaged, they have to live with the consequences. If a female's X is damaged, she can often ignore the consequences. This represents to date one of the strongest pieces of evidence showing the involvement of X chromosomes in brain function and thus brain behavior.

Mental health professionals have known for years about sex-based differences in the type and severity of psychiatric disorders. Males are more severely afflicted by schizophrenia than females, for example. By more than 2 to 1, women are more likely to get depressed than men, a figure that shows up just after puberty and remains stable for the next 50 *years*. Males exhibit more antisocial behavior. Females have more anxiety. Most alcoholics and drug addicts are male. Most anorexics are female. Says Thomas Insel, from the National Institute of Mental Health, "It's pretty difficult to find any single factor that's more predictive for some of these disorders than gender."

But what about normal behavior? The three research islands have very few bridges between them. There are bridge-construction projects, however, and we are going to talk about two of the best.

dealing with traumatic situations

It's a horrible slideshow. In it, a little boy is run over by a car while walking with his parents. If you ever see that show, you will never forget it. But what if you *could* forget it? The brain's amygdala aids in the creation of emotions and our ability to remember them. Suppose there was a magic elixir that could momentarily suppress it? Such an elixir does exist, and it was used to show that men and women process emotions differently.

You have probably heard the term left brain vs. right brain. You may have heard that this underscores creative vs. analytical people. That's a folk tale, the equivalent of saying the left side of a luxury liner is responsible for keeping the ship afloat, and the right is responsible for making it move through the water. Both sides are involved in both processes. That doesn't mean the hemispheres are equal, however. The right side of the brain tends to remember the gist of an experience, and the left brain tends to remember the details.

Researcher Larry Cahill eavesdropped on men's and women's brains under acute stress (he showed them slasher films), and what he found is this: Men handled the experience by firing up the amygdala in their brain's right hemisphere. Their left was comparatively silent. Women handled the experience with the opposite hemisphere. They lit up their left amygdala, their right comparatively silent. If males are firing up the right hemisphere (the "gist dictator"), does that mean males remember more gist than detail of a given emotional experience related to stress? Do females remember more detail than gist of an emotional experience related to stress? Cahill decided to find out.

That magic elixir of forgetting, a drug called propranolol, normally is used to regulate blood pressure. As a beta-blocker, it also inhibits the biochemistry that usually would activate the amygdala during emotional experiences. The drug is being investigated as a potential treatment for combat-related disorders.

But Cahill gave it to his subjects before they watched a traumatic film. One week later, he tested their memories of it. Sure enough, the men lost the ability to recall the gist of the story, compared with men who didn't take the drug. Women lost the ability to recall the details. One must be careful not to overinterpret these data. The results clearly define only emotional responses to stressful situations, not objective details and summaries. This is not a battle between the accountants and the visionaries.

Cahill's results come on the heels of similar findings around the world. Other labs have extended his work, finding that women recall more emotional autobiographical events, more rapidly and with greater intensity, than men do. Women consistently report more vivid memories for emotionally important events such as a recent argument, a first date, or a vacation. Other studies show that, under stress, women tend to focus on nurturing their offspring, while men tend to withdraw. This tendency in females has sometimes been called "tend and befriend." Its origins are unknown, and the reason comes straight from the mouth of Stephen Jay Gould: "It is logically, mathematically, and scientifically impossible to pull them apart."

This quote reminds me of my two sons in a fight, but Gould is actually talking about the age-old nature vs. nurture argument.

verbal communication

Behaviorist Deborah Tannen has done some fascinating work in this area, studying gender differences in verbal capacity. The Cliff Notes version of Tannen's and others' findings over the past 30 years: "Women are better at it." Though the specifics are often controversial, much of the empirical support comes from unusual quarters, including brain pathologies. We have known for years that language and reading disorders occur approximately twice as often in little boys as in little girls. Women also recover from stroke-induced verbal impairment better than men. Many researchers suspect that risk disparities like these hint at underlying differences in normal cognition. They often point to neuroanatomical data to explain the difference: Women tend to use both hemispheres when speaking and processing verbal information. Men primarily use one. Women tend to have thick cables connecting their two hemispheres. Men's are thinner. It's as though females have a backup system that is absent in males.

These clinical data have been used to support findings first noticed by educators. Girls seem verbally more sophisticated than little boys as they go through the school system. They are better at verbal memory tasks, verbal fluency tasks, and speed of articulation. When these little girls grow up, they are still champions at processing verbal information. Real as these data seem, however, almost none of them can be divorced from a social context. That's why Gould's comment is so helpful.

Tannen spent a long time observing and videotaping how little girls and little boys interact with each other. Her original question was to find out how boys and girls of different ages talked to their best friends, and if any detectable patterns emerged. If she found some, she wanted to know how stable they were. Would the patterns detected in childhood also show up in college students? The patterns she found were predictable and stable, independent of age and geography. The conversational styles we've developed as adults come directly from the same-sex interactions we solidified as children. Tannen's findings center on three areas.

cementing relationships

When girl best friends communicate with each other, they lean in, maintain eye contact, and do a lot of talking. They use their sophisticated verbal talents to cement their relationships. Boys never do this. They rarely face each other directly, preferring either parallel or oblique angles. They make little eye contact, their gaze always casting about the room. They do not use verbal information to cement their relationships. Instead, commotion seems to be the central currency of a little boy's social economy. Doing things physically together is the glue that holds their relationships intact.

My sons, Josh and Noah, have been playing a one-upmanship game since they were toddlers. A typical version might involve ball throwing. Josh would say, "I can throw this up to the ceiling," and would promptly do so. Then they would laugh. Noah would respond by grabbing the ball, saying, "Oh yeah? I can throw this up to the sky," and throwing the ball higher. This ratcheting, with laughter, would continue until they reached the "galaxy" or the big prize, "God."

Tannen saw this consistent style everywhere she looked—except when observing little girls. The female version goes something like this. One sister says, "I can take this ball and throw it to the ceiling," and she promptly does. She and her sibling both laugh. The other sister grabs the ball, throws it up to the ceiling, and says, "I can, too!" Then they talk about how cool it is that they can both throw the ball at the same height. This style persists into adulthood for both sexes.

Tannen's data, unfortunately, have been misinterpreted as "Boys always compete, and girls always cooperate." As this example shows, however, boys are being extremely cooperative. They are simply doing it through competition, deploying their favorite strategy of physical activity.

negotiating status

By elementary school, boys finally start using their verbal skills for something: to negotiate their status in a large group. Tannen found that high-status males give orders to the rest of the group, verbally or even physically pushing the low-status boys around. The "leaders" maintain their fiefdoms not only by issuing orders but by making sure the orders are carried out. Other strong members try to challenge them, so the guys at the top learn quickly to deflect challenges. This is often done with words as well. The upshot is that the hierarchy is very evident with boys. And hard. The life of a low-status male is often miserable. Independent behavior, which is a characteristic of control at the top, tends to be highly prized.

Tannen found very different behaviors when observing little girls. There were both high-status and low-status females, as with the boys. But they used strikingly different strategies to generate and maintain their hierarchies. The girls spend a lot of time talking. This communication is so important that the type of talk determines the status of the relationship. To whom you tell your secrets determines "best friend" status. The more secrets revealed, the more likely the girls identify each other as close. Girls tend to de-emphasize the status between them in these situations. Using their sophisticated verbal ability, the girls tend not to give top-down imperial orders. If one of the girls tries issuing commands, the style is usually rejected: The girl is tagged as "bossy" and isolated socially. Not that decisions aren't made. Various members of the group give suggestions, then discuss alternatives. Eventually, a consensus emerges.

The difference between the genders could be described as the addition of a single powerful word. Boys might say, "Do this." Girls would say, "*Let's* do this."

into adulthood

Tannen found that over time, these ways of using language became increasingly reinforced, which incited different social sensitivities in the two groups. Any boy who gave orders was a leader. Any girl who gave orders was bossy. By college age, most of these styles were deeply entrenched. And that's when the problems became most noticeable, showing up at work and in marriage.

A 20-something newlywed was on a drive with her girlfriend, Emily. She became thirsty. "Emily, are you thirsty?" she asked. With lifelong experience at verbal negotiation, Emily knew what her friend wanted. "I don't know. Are *you* thirsty?" she responded. There then ensued a small discussion about whether they were both thirsty enough to stop the car and get water.

A few days later, the woman was driving with her husband. "Are you thirsty?" she asked. "No, I'm not," he replied. They actually got into an argument that day. She was annoyed because she had wanted to stop; he was annoyed because she wasn't direct. This type of conflict would become increasingly familiar as their marriage aged.

Such scenarios can play out in the work force just as easily. Women who exert "male" leadership styles are in danger of being perceived as bossy. Men who do the same thing are often praised as decisive. Tannen's great contribution was to show that these stereotypes form very early in our social development, perhaps assisted by asymmetric verbal development. They transcend geography, age, and even time. Tannen, who was an English literature major, sees these tendencies in manuscripts that go back centuries.

nature or nurture?

Tannen's findings are statistical patterns, not an all-or-none phenomenon. She has found that many factors affect our language patterns. Regional background, individual personality, profession, social class, age, ethnicity, and birth order all affect how we use language to negotiate our social ecologies. Boys and girls are treated differently socially the moment they are born, and they are often reared in societies filled with centuries of entrenched prejudice. It would be a miracle if we somehow transcended our experience and behaved in an egalitarian fashion.

Given the influence of culture on behavior, it is overly simplistic to invoke a purely biological explanation for Tannen's observations. And, given the great influence of brain biology on behavior, it is also simplistic to invoke a purely social explanation. The real answer to the nature-or-nurture question is "We don't know." That can be frustrating to hear. Everybody wants to build bridges between these islands. Cahill, Tannen, and countless others are doing their best to provide us with the boards and nails. That's not the same thing as saying the connections exist, however. Believing that there are strong associations between genes and cells and behaviors when there are none is not only wrong but dangerous. Just ask Larry Summers.

ideas

How can we use these data in the real world?

Get the facts straight on emotions

Dealing with the emotional lives of men and women is a big part of the job for teachers and business professionals. They need to know:

1) Emotions are useful. They make the brain pay attention.

2) Men and women process certain emotions differently.

3) The differences are a product of complex interactions between nature and nurture.

Try different gender arrangements in the classroom

My son's third-grade teacher began seeing a stereotype that

worsened as the year progressed. The girls were excelling in the language arts, and the boys were pulling ahead in math and science. This was only the *third grade!* The language-arts differences made some sense to her. But she knew there was no statistical support for the contention that men have a better aptitude for math and science than women. Why, for heaven's sake, was she presiding over a stereotype?

The teacher guessed that part of the answer lay in the students' social participation during class. When the teacher asked a question of the class, who answered first turned out to be unbelievably important. In the language arts, the girls invariably answered first. Other girls reacted with that participatory, "me too" instinct. The reaction on the part of the boys was hierarchical. The girls usually knew the answers, the boys usually did not, and the males responded by doing what low-status males tend to do: They withdrew. A performance gap quickly emerged. In math and science, boys and girls were equally likely to answer a question first. But the boys used their familiar "top each other" conversational styles when they participated, attempting to establish a hierarchy based on knowledge aptitude. This included drubbing anyone who didn't make the top, including the girls. Bewildered, the girls began withdrawing from participating in the subjects. Once again, a performance gap emerged.

The teacher called a meeting of the girls and verified her observations. Then she asked for a consensus about what they should do. The girls decided that they wanted to learn math and science separately from the boys. Previously a strong advocate for mixed-gender classes, the teacher wondered aloud if that made any sense. Yet if the girls started losing the math-and-science battle in the third grade, the teacher reasoned they were not likely to excel in the coming years. She obliged. It took only two weeks to close the performance gap.

Can the teacher's result be applied to classrooms all over the world? Actually, the experiment is not a result at all. It is a comment. This is not a battle that can be won by testing one classroom in a single school year. This is a battle properly fought by testing hundreds of classrooms and thousands of students from all walks of life, over a period of years.

Use gender teams in the workplace

One day, I spoke about gender with a group of executives-in-training at the Boeing Leadership Center in St. Louis. After showing some of Larry Cahill's data about gist and detail, I said, "Sometimes women are accused of being more emotional than men, from the home to the workplace. I think that women might not be any more emotional than anyone else." I explained that because women perceive their emotional landscape with more data points (that's the detail) and see it in greater resolution, women may simply have more information to which they are capable of reacting. If men perceived the same number of data points, they might have the same number of reactions. Two women in the back began crying softly. After the lecture, I asked them about their reactions, fearing I may have offended them. What they said instead blew me away. "It was the first time in my professional life," one of them said, "that I didn't feel like I had to apologize for who I was."

And that got me to thinking. In our evolutionary history, having a team that could simultaneously understand the gist and details of a given stressful situation helped us conquer the world. Why would the world of business be exempted from that advantage? Having an executive team or work group capable of simultaneously understanding both the emotional forests and the trees of a stressful project, such as a merger, might be a marriage made in business heaven. It could even affect the bottom line.

Companies often conduct management training with situation simulations. They could take a mixed-gendered team and a unisex team and have them go at a project together. Take another set of two teams, but first teach them about the known gender-based differences before taking on the same project. You have four potential outcomes. Would the mixed teams do better than the mono teams? Would the trained groups do better than the untrained groups? Would these results be stable in, say, six months? You might find that management teams with a gist/detail balance create the best shot for productivity. At the very least, this means that both men and women have an equal right to be at the decision-making table.

We could have environments where gender differences are both noted and celebrated, as opposed to ignored and marginalized. Had this been done earlier, we might have more women in science and engineering now. We might have shattered the archetypal glass ceiling and saved companies a lot of money. Heck, it may even have salvaged the Harvard president's job.

Summary

Rule #11
Male and female brains are different.

* The X chromosome that males have one of and females have two of—though one acts as a backup—is a cognitive "hot spot," carrying an unusually large percentage of genes involved in brain manufacture.

* Women are genetically more complex, because the active X chromosomes in their cells are a mix of Mom's and Dad's. Men's X chromosomes all come from Mom, and their Y chromosome carries less than 100 genes, compared with about 1,500 for the X chromosome.

* Men's and women's brains are different structurally and biochemically—men have a bigger amygdala and produce serotonin faster, for example—but we don't know if those differences have significance.

* Men and women respond differently to acute stress: Women activate the left hemisphere's amygdala and remember the emotional details. Men use the right amygdala and get the gist.

Get more at www.brainrules.net

exploration

Rule #12
We are powerful and natural explorers.

MY DEAR SON JOSH got a painful bee sting at the tender age of 2, and he almost deserved it.

It was a warm, sunny afternoon. We were playing the "pointing game," a simple exercise where he would point at something, and I would look. Then we'd both laugh. Josh had been told not to touch bumblebees because they could sting him; we used the word "danger" whenever he approached one. There, in a patch of clover, he spotted a big, furry, buzzing temptress. As he reached for it, I calmly said, "Danger," and he obediently withdrew his hand. He pointed at a distant bush, continuing our game.

As I looked toward the bush, I suddenly heard a 110-decibel yelp. While I was looking away, Josh reached for the bee, which promptly stung him. Josh had used the pointing game as a diversion, and I was outwitted by a 2-year-old.

"DANGER!" he sobbed as I held him close.

"Danger," I repeated sadly, hugging him, getting some ice, and wondering what puberty would be like in 10 years or so.

This incident was Dad's inauguration into a behavioral suite often called the terrible twos. It was a rough baptism for me and the little guy. Yet it also made me smile. The mental faculties kids use to distract their dads are the same they will use as grownups to discover the composition of distant suns or the next alternative energy. We are natural explorers, even if the habit sometimes stings us. The tendency is so strong, it is capable of turning us into lifelong learners. But you can see it best in our youngest citizens (often when they seem at their worst).

breaking stuff

Babies give researchers a clear view, unobstructed by years of contaminating experiences, of how humans naturally acquire information. Preloaded with lots of information-processing software, infants acquire information using surprisingly specific strategies, many of which are preserved into adulthood. In part, understanding how humans learn at this age means understanding how humans learn at any age.

We didn't always think that way. If you had said something about preset brain wiring to researchers 40 years ago, their response would have been an indignant, "What are you smoking?" or, less politely, "Get out of my laboratory." This is because researchers for decades thought that babies were a blank slate—a tabula rasa. They thought that everything a baby knew was learned by interactions with its environments, primarily with adults. This perspective undoubtedly was formulated by overworked scientists who never had any children. We know better now. Amazing strides have been made in understanding the cognitive world of the infant. Indeed, the research world now looks to babies to show how humans, including adults, think about practically everything.

Let's look under the hood of an infant's mind at the engine that drives its thinking processes and the motivating fuel that keeps its intellect running.

This fuel consists of a clear, high-octane, unquenchable need to know. Babies are born with a deep desire to understand the world around them and an incessant curiosity that compels them to aggressively explore it. This need for explanation is so powerfully stitched into their experience that some scientists describe it as a drive, just as hunger and thirst and sex are drives.

Babies seem preoccupied by the physical properties of objects. Babies younger than a year old will systematically analyze an object with every sensory weapon at their disposal. They will feel it, kick it, try to tear it apart, stick it in their ear, stick it in their mouth, give it to you so you can stick it in your mouth. They appear to be intensely gathering information about the properties of the object. Babies methodically do experiments on the objects to see what else they will do. In our household, this usually meant breaking stuff.

These object-oriented research projects grow increasingly sophisticated. In one famous set of experiments, babies were given a rake and a toy set far apart from each other. The babies quickly learned to use the rake to get the toy. This is not exactly a groundbreaking discovery, as every parent knows. Then the researchers observed an astonishing thing. After a few successful attempts, the babies lost interest in the toy. But not in the experiment. They would take the toy and move it to different places, then use the rake to grab it. They even placed the toy out of reach to see what the rake could do. The toy didn't seem to matter to them at all. What mattered was the fact that the rake could move it closer. They were experimenting with the relationship between objects, specifically with how one object could influence the other.

Hypothesis testing like that is the way all babies gather information. They use a series of increasingly self-corrected ideas to figure out how the world works. They actively test their environment, much as a scientist would: Make a sensory observation, form a hypothesis about what is going on, design an experiment capable of testing the hypothesis, and then draw conclusions from the findings.

tongue testing

In 1979, Andy Meltzoff rocked the world of infant psychology by sticking out his tongue at a newborn and being polite enough to wait for a reply. What he found astonished him. The baby stuck her tongue back out at him! He reliably measured this imitative behavior with infants only 42 minutes old. The baby had never seen a tongue before, not Meltzoff's and not her own, yet the baby knew she had a tongue, knew Meltzoff had a tongue, and somehow intuited the idea of mirroring. Further, the baby knew that if she stimulated a series of nerves in a certain sequence, she could also stick her tongue out (definitely not consistent with the notion of tabula rasa).

I tried this with my son Noah. He and I started our relationship in life by sticking our tongues out at each other. In his first 30 minutes of life, we had struck up an imitative conversation. By the end of his first week, we were well entrenched in dialogue: Every time I came into his crib room, we greeted each other with tongue protrusions. It was purely adaptive on his part, even as it was purely delightful on my part. If I had not stuck my tongue out initially, he would not be doing so with such predictability every time I came into his visual range.

Three months later, my wife picked me up after a lecture at a medical school, Noah in tow. I was still fielding questions, but I scooped up Noah and held him close while answering. Out of the corner of my eye, I noticed Noah gazing at me expectantly, flicking his tongue out about every five seconds. I smiled and stuck my tongue out at Noah mid-question. Immediately he squealed and started sticking his tongue out with abandon, every half-second or so. I knew exactly what he was doing. Noah made an observation (Dad and I stick our tongues out at each other), formed a hypothesis (I bet if I stick my tongue out at Dad, he will stick his tongue back out at me), created and executed his experiment (I will stick my tongue out at Dad), and changed his behavior as a result of the evaluation of his research (sticking his tongue out more frequently). Nobody taught Noah, or any other baby, how to do this. And it is a lifelong strategy. You probably did it this morning when you couldn't find your glasses, hypothesized they were in the laundry room, and went downstairs to look. From a brain science perspective, we don't even have a good metaphor to describe how you know to do that. It is so automatic, you probably had no idea you were looking at the results of a successful experiment when you found your spectacles lying on the dryer.

Noah's story is just one example of how babies use their precious preloaded information-gathering strategies to gain knowledge they didn't have at birth. We also can see it in disappearing cups and temper tantrums.

Little Emily, before 18 months of age, still believes that if an object is hidden from view, that object has disappeared. She does not have what is known as "object permanence." That is about to change. Emily has been playing with a washcloth and a cup. She covers the cup with the cloth, and then pauses for a second, a concerned look on her brow. Slowly she pulls the cloth away from the cup. The cup is still there! She glares for a moment, then quickly covers it back up. Thirty seconds go by before her hand tentatively reaches for the cloth. Repeating the experiment, she slowly removes the cloth. The cup is *still* there! She squeals with delight. Now things go quickly. She covers and uncovers the cup again and again, laughing loudly each time. It is dawning on Emily that the cup has object permanence: Even if removed from view, it has not disappeared. She will repeat this experiment for more than half an hour. If you have ever spent time with an 18-month old, you know that getting one to concentrate on anything for 30 minutes is some kind of miracle. Yet it happens, and to babies at this age all over the world.

Though this may sound like a delightful form of peek-a-boo, it is actually an experiment whose failure would have lethal evolutionary consequences. Object permanence is an important concept to have if you live in the savannah. Saber-toothed tigers still exist, for example, even if they suddenly duck down in the tall grass. Those who didn't acquire this knowledge usually were on some predator's menu.

testing you, too

The distance between 14 months of age and 18 months of age is extraordinary. This is when children begin to learn that people have desires and preferences separate from their own. They don't start out that way. They think that because they like something, the whole world likes the same thing. This may be the origin of the "Toddler's Creed," or what I like to call "Seven Rules of Management from a Baby's Perspective":

If I want it, it is mine.
If I give it to you and change my mind later, it is mine.
If I can take it away from you, it is mine.
If we are building something together, all of the pieces are mine.
If it looks just like mine, it is mine.
If it is mine, it will never belong to anybody else, no matter what.
If it is yours, it is mine.

At 18 months, it dawns on babies that this viewpoint may not always be accurate. They begin to learn that adage that most newlyweds have to relearn in spades: "What is obvious to you is obvious to you."

How do babies react to such new information? By testing it, as usual. Before the age of 2, babies do plenty of things parents would rather them not do. But after the age of 2, small children will do things *because* their parents don't want them to. The compliant little darlings seem to transform into rebellious little tyrants. Many parents think their children are actively defying them at this stage. (The thought certainly crossed my mind as I nursed Joshua's unfortunate bee sting.) That would be a mistake, however. This stage is simply the natural extension of a sophisticated research program begun at birth.

You push the boundaries of people's preferences, then stand back and see how they react. Then you repeat the experiment, pushing them to their limits over and over again to see how stable the findings are, as if you were playing peek-a-boo. Slowly you begin to perceive the length and height and breadth of people's desires, and how they differ from yours. Then, just to be sure the boundaries are still in place, you occasionally do the whole experiment over again.

Babies may not have a whole lot of understanding about their world, but they know a whole lot about how to get it. It reminds me of the old Chinese proverb "Catch me a fish and I eat for a day; teach me to fish and I eat for a lifetime."

monkey see, monkey do

Why does a baby stick its tongue back out at you? The beginnings of a neural road map have been drawn in the past few years, at least for some of the "simpler" thinking behaviors, such as imitation. Three investigators at the University of Parma were studying the macaque, assessing brain activity as it reached for different objects in the laboratory. The researchers recorded the pattern of neural firing when the monkey picked up a raisin. One day, researcher Leonardo Fogassi walked into the laboratory and casually plucked a raisin from a bowl. Suddenly, the monkey's brain began to fire excitedly. The recordings were in the raisin-specific pattern, *as if the animal had just picked up the raisin*. But the monkey had not picked up the raisin. It simply saw Fogassi do it.

The astonished researchers quickly replicated and extended their findings, and then published them in a series of landmark papers describing the existence of "mirror neurons." Mirror neurons are cells whose activity reflect their surroundings. Cues that could elicit mirror neural responses were found to be remarkably subtle. If a primate simply heard the sound of someone doing something it had previously experienced—say, tearing a piece of paper—these neurons could fire as if the monkey were experiencing the full stimulus. It wasn't long before researchers identified human mirror neurons. These neurons are scattered across the brain, and a subset is involved in action recognition—that classic imitative behavior such as babies sticking out their tongues. Other neurons mirror a variety of motor behaviors.

We also are beginning to understand which regions of the brain are involved in our ability to learn from a series of increasingly self-corrected ideas. We use our right prefrontal cortex to predict error and to retrospectively evaluate input for errors. The anterior cingulate cortex, just south of the prefrontal cortex, signals us when perceived unfavorable circumstances call for a change in behavior. Every year, the brain reveals more and more of its secrets, with babies leading the way.

a lifetime journey

We do not outgrow the thirst for knowledge, a fact brought home to me as a post-doc at the University of Washington. In 1992, Edmond Fischer shared with Edwin Krebs the Nobel Prize in Physiology or Medicine. I had the good fortune to be familiar with both their work and their offices. They were just down the hall from mine. By the time I arrived, they were already in their mid-70s. The first thing I noticed upon meeting them was that they were not retired. Not physically and not mentally. Long after they had earned the right to be put out to scientific pasture, both had powerful, productive laboratories in full swing. Every day I would see them walking down the hall, oblivious to others, chatting about some new finding, swapping each other's journals, listening intently to each other's ideas. Sometimes they would have someone else along, grilling them and in turn being grilled about some experimental result. They were creative like artists, wise as Solomon, lively as children. They had lost *nothing*. Their intellectual engines were still revving, and curiosity remained the fuel. That's because our learning abilities don't have to change as we age. We can remain lifelong learners.

There may have been strong evolutionary pressure for maintaining these strategies. Problem-solving was greatly favored in the unstable environment of the Serengeti. But not just any kind of problem-solving. When we came down from the trees to the savannah, we did not say to ourselves, "Good lord, give me a book and a lecture and a board of directors so I can spend 10 years learning how to survive in this place." Our survival did not depend upon exposing ourselves to organized, pre-planned packets of information. Our survival depended upon chaotic, reactive information-gathering experiences. That's why one of our best attributes is the ability to learn through a series of increasingly self-corrected ideas. "The red snake with the white stripe bit me yesterday, and I almost died," is an observation we readily made. Then we went a step further: "I hypothesize that if I encounter the same snake, the same thing will happen!" It is a scientific learning style we have explored literally for millions of years. It is not possible to outgrow it in the whisper-short seven to eight decades we have on the planet.

Researchers have shown that some regions of the adult brain stay as malleable as a baby's brain, so we can grow new connections, strengthen existing connections, and even create new neurons, allowing all of us to be lifelong learners. We didn't always think that. Until five or six years ago, the prevailing notion was that we were born with all of the brain cells we were ever going to get, and they steadily eroded in a depressing journey through adulthood to old age. We do lose synaptic connections with age (some estimates of neural loss alone are close to 30,000 neurons per day). But the adult brain also continues creating neurons within the regions normally involved in learning. These new neurons show the same plasticity as those of newborns. The adult brain throughout life retains the ability to change its structure and function in response to experience.

Can we continue to explore our world as we age? I can almost

271

hear Krebs and Fischer saying, "Well, duh. Next question." Of course, we don't always find ourselves in environments that encourage such curiosity as we grow older. I've been fortunate to have a career that allowed me the freedom to pick my own projects. Before that, I was lucky to have my mother.

from dinosaurs to atheism

I remember, when I was 3 years old, obtaining a sudden interest in dinosaurs. I had no idea that my mother had been waiting for it. That very day, the house began its transformation into all things Jurassic. And Triassic. And Cretaceous. Pictures of dinosaurs would go up on the wall. I would begin to find books about dinosaurs strewn on the floor and sofas. Mom would even couch dinner as "dinosaur food," and we would spend hours laughing our heads off trying to make dinosaur sounds. And then, suddenly, I would lose interest in dinosaurs, because some friend at school acquired an interest in spaceships and rockets and galaxies. Extraordinarily, my mother was waiting. Just as quickly as my whim changed, the house would begin its transformation from big dinosaurs to Big Bang. The reptilian posters came down, and in their places, planets would begin to hang from the walls. I would find little pictures of satellites in the bathroom. Mom even got "space coins" from bags of potato chips, and I eventually gathered all of them into a collector's book.

This happened over and over again in my childhood. I got an interest in Greek mythology, and she transformed the house into Mount Olympus. My interests careened into geometry, and the house became Euclidean, then cubist. Rocks, airplanes. By the time I was 8 or 9, I was creating my own house transformations.

One day, around age 14, I declared to my mother that I was an atheist. She was a devoutly religious person, and I thought this announcement would crush her. Instead, she said something like "That's nice, dear," as if I had just declared I no longer liked nachos. The next day, she sat me down by the kitchen table, a wrapped package in her lap. She said calmly, "So, I hear you are now an atheist. Is that true?" I nodded yes, and she smiled. She placed the package in my hands. "The man's name is Friedrich Nietzsche, and the book is called *Twilight of the Idols*," she said. "If you are going to be an atheist, be the best one out there. *Bon appetit!*"

I was stunned. But I understood a powerful message: Curiosity itself was the most important thing. And what I was interested in *mattered*. I have never been able to turn off this fire hose of curiosity.

Most developmental psychologists believe that a child's need to know is a drive as pure as a diamond and as distracting as chocolate. Even though there is no agreed-upon definition of curiosity in cognitive neuroscience, I couldn't agree more. I firmly believe that if children are allowed to remain curious, they will continue to deploy their natural tendencies to discover and explore until they are 101. This is something my mother seemed to know instinctively.

For little ones, discovery brings joy. Like an addictive drug, exploration creates the need for more discovery so that more joy can be experienced. It is a straight-up reward system that, if allowed to flourish, will continue into the school years. As children get older, they find that learning not only brings them joy, but it also brings them mastery. Expertise in specific subjects breeds the confidence to take intellectual risks. If these kids don't end up in the emergency room, they may end up with a Nobel Prize.

I believe it is possible to break this cycle, anesthetizing both the process and the child. By first grade, for example, children learn that education means an A. They begin to understand that they can acquire knowledge not because it is interesting, but because it can get them something. Fascination can become secondary to "What do I need to know to get the grade?" But I also believe the curiosity instinct is so powerful that some people overcome society's message to go to sleep intellectually, and they flourish anyway.

My grandfather was one of those people. He was born in 1892 and lived to be 101 years old. He spoke eight languages, went through several fortunes, and remained in his own house (mowing his own lawn) until the age of 100, lively as a firecracker to the end. At a party celebrating his centenary, he took me aside. "You know, *Juanito*," he said, clearing his throat, "sixty-six years separate the Wright brothers' airplane from Neil Armstrong and the moon." He shook his head, marveling. "I was born with the horse and buggy. I die with the space shuttle. What kind of thing is that?" His eyes twinkled. "I live the good life!"

He died a year later.

I think of him a lot when I think of exploration. I think of my mother and her magically transforming rooms. I think of my youngest son experimenting with his tongue, and my oldest son's overwhelming urge to take on a bee sting. And I think that we must do a better job of encouraging lifelong curiosity, in our workplaces and especially in our schools.

ideas

Google takes to heart the power of exploration. For 20 percent of their time, employees may go where their mind asks them to go. The proof is in the bottom line: Fully 50 percent of new products, including Gmail and Google News, came from "20 percent time." How would we implement such freedom in classrooms? Some people have tried to harness our natural exploratory tendencies by using "problem-based" or "discovery-based" learning models. These models have both strong advocates and strong detractors. Most agree that these debates are missing hard-nosed empirical results that show the long-term effects of these styles. I would go further and argue that what is missing is a real live laboratory in which brain scientists and education scientists could carry out investigations on a routine, long-term basis. I would like to describe the place for such research.

Analyze the success of medical schools

In the early 20th century, John Dewey created a laboratory school at the University of Chicago, in part because he thought that learning should be tested in real-world situations. Though such schools fell out of favor in the mid-'60s, perhaps with good reason, a 21ˢᵗ-century version might look to one of the most successful educational models out there, a medical school. As William H. Payne, a colleague of Dewey's, said, "Psychology, in fact, stands in the same relation to teaching that anatomy does to medicine." It still does, though I would replace "psychology" with "brain science."

The best medical-school model has three components: a teaching hospital; faculty who work in the field as well as teach; and research laboratories. It is a surprisingly successful way of treating people. It is also a surprisingly successful way to transfer complex information from one brain to another. I often have watched bright non-science majors become accepted into a medical-school program and then, within four years, transform into gifted healers and terrific scientists.

Why do you get good health and good training at the same time? I am convinced that it is the structure.

1) Consistent exposure to the real world

By combining traditional book-learning and a teaching hospital, the student gets an unobstructed view of what they are getting into *while they are going through it.* Most medical students stroll through a working hospital on their way to class every day of their training lives. They confront on a regular basis the very reason they chose medical school in the first place. By the third year, most students are in class only half of the time. They spend the other half learning on the job in the teaching hospital or a clinic associated with it. Residencies come next for more real-world experience.

2) Consistent exposure to people who operate in the real world

Medical students are taught by people who actually do what they teach as their "day job." In more recent years, these people are not only practicing medical doctors, but practicing medical researchers involved in cutting-edge projects with powerful clinical implications. Medical students are asked to participate.

3) Consistent exposure to practical research programs

Here's a typical experience: The clinician-professor is lecturing in a traditional classroom setting and brings in a patient to illustrate some of his points. The professor announces: "Here is the patient. Notice that he has disease X with symptoms A, B, C, and D." He then begins to lecture on the biology of disease X. While everybody is taking notes, a smart medical student raises her hand and says, "I see symptoms A, B, and C. What about symptoms E, F, and G?" The professor looks a bit chagrined (or excited) and responds, "We don't know about symptoms E, F, and G." You can hear a pin drop at those moments, and the impatient voices whispering inside the students' heads are almost audible: "Well, let's find out!" These are the opening words of most of the great research ideas in human medicine.

That's true exploratory magic. By simple juxtaposition of real-world needs with traditional book learning, a research program is born. The tendency is so strong that you have to deliberately cut off the discussions to keep the ideas from forming. Most programs have chosen not to cut off such discussions. As a result, most American medical schools possess powerful research wings.

This model gives students a rich view of the field of medicine. Not only are they taught by people who are involved in the day-to-day aspect of healing, but they are exposed to people who are trained to think about the future of medicine. These scientists represent the brightest minds in the country. And this model provides the single most natural harness for the exploratory instincts of the human species I have ever encountered.

Create a college of education that studies the brain

I envision a college of education where the program is all about brain development. It is divided into three parts, like a medical school. It has traditional classrooms. It is a community school staffed and run by three types of faculty: traditional education faculty, certified teachers who teach the little ones, and brain scientists. This last group teaches in research labs devoted to a single purpose: investigating how the human brain learns in teaching environments, then actively testing hypothesized ideas in real-world classroom situations.

Students would get a Bachelor of *Science* in education. The future educator is infused with deep knowledge about how the human brain acquires information. Topics range from structural brain anatomy to psychology, from molecular biology to the latest in cognitive neuroscience. But the coursework is only the beginning. After their first year of study, students would start actively participating in the life of the onsite school.

One semester might be devoted to understanding the development of the teenage brain. The internship would involve assisting in a junior high and high school. Another semester might be devoted to behavioral pathologies such as attention deficit hyperactivity disorder, and students would assist in a special-education class. Still another course would be devoted to the effects of family life on human learning, with students attending parent association meetings and observing parent-teacher conferences. In this two-way interaction, the insights of researchers and the insights of practitioners have a chance to marinate in a single ongoing intellectual environment. The model creates a vigorous, use-driven strategic research-and-development program. The practitioner is elevated to the role of colleague, an active partner in helping shape the research direction, even as the researcher helps the practitioner form the specifics of the effort.

This model honors our evolutionary need to explore. It creates teachers who know about brain development. And it's a place to do the real-world research so sorely needed to figure out how, exactly, the rules of the brain should be applied to our lives. The model could be imported to other academic subjects as well. A business school teaching how to run a small business might actually run one, for example, as a part of its academic life.

the sense of wonder

If you could step back in time to one of the first real Western-style universities, say, the University of Bologna, and visit its biology labs, you would laugh out loud. I would join you. By today's standards, biological science in the 11th century was a joke, a strange mix of astrological influences, religious forces, dead animals, and rudely smelling chemical concoctions, some of which were toxic.

But if you went down the hall and peered inside Bologna's standard lecture room, you wouldn't feel as if you were in a museum. You would feel at home. There is a lectern for the teacher to hold forth, surrounded by chairs where students absorb whatever is being held forth. Minus perhaps an overhead or two, it looks remarkably similar to today's classrooms. Could it be time for a change?

My sons most likely would say yes. They and my mother are probably the greatest teachers I ever had.

My 2-year-old son Noah and I were walking down the street on our way to preschool when he suddenly noticed a shiny pebble embedded in the concrete. Stopping midstride, the little guy considered it for a second, found it thoroughly delightful, and let out a laugh. He spied a small plant an inch farther, a weed valiantly struggling through a crack in the asphalt. He touched it gently, then laughed again. Noah noticed beyond it a platoon of ants marching in single file, which he bent down to examine closely. They were carrying a dead bug, and Noah clapped his hands in wonder. There were dust particles, a rusted screw, a shiny spot of oil. Fifteen minutes had passed, and we had gone only 20 feet. I tried to get him to move along, having the audacity to act like an adult with a schedule. He was having none of it. And I stopped, watching my little teacher, wondering how long it had been since I had taken 15 minutes to walk 20 feet.

The greatest Brain Rule of all is something I cannot prove or characterize, but I believe in it with all my heart. As my son was trying to tell me, it is the importance of curiosity.

For his sake and ours, I wish classrooms and businesses were designed with the brain in mind. If we started over, curiosity would be the most vital part of both demolition crew and reconstruction crew. As I hope to have related here, I am very much in favor of both.

I will never forget the moment this little professor taught his daddy about what it meant to be a student. I was thankful and a little embarrassed. After 47 years, I was finally learning how to walk down the street.

Summary

Rule #12
We are powerful
and natural explorers.

* Babies are the model of how we learn—not by passive reaction to the environment but by active testing through observation, hypothesis, experiment, and conclusion.

* Specific parts of the brain allow this scientific approach. The right prefrontal cortex looks for errors in our hypothesis ("The saber-toothed tiger is not harmless"), and an adjoining region tells us to change behavior ("Run!").

* We can recognize and imitate behavior because of "mirror neurons" scattered across the brain.

* Some parts of our adult brains stay as malleable as a baby's, so we can create neurons and learn new things throughout our lives.

Get more at www.brainrules.net

references

*Extensive, notated references
and illustrations are online
at www.brainrules.net*

acknowledgements

IN A LIST OF just about anything, items at the beginning and end are the easiest for the brain to retrieve. It's called Serial Position Effect, and I mention it because I am about to list some of the many people who helped bring this project to fruition. There obviously will be a first person and a last person and lots of people in between. This is not because I see these folks in a hierarchy of values; it is simply because written languages are necessarily, cursedly linear. Please pay attention, dear reader, to the folks in the middle as well as to those at the end points. As I have often mentioned to graduate students, there is great value in the middle of most U-shaped curves.

First, I thank my publisher at Pear Press, Mark Pearson, the guiding hand of this project and easily the wisest, oldest young man with whom I have ever had the joy to work. It was a pleasure to work with editor Tracy Cutchlow, who with patience, laughter, and extraordinary thoughtfulness, taught me how to write.

Special thanks to Dan Storm and Eric Chudler for providing invaluable scientific comments and expertise.

I am grateful to friends on this journey with me: Lee Huntsman, for hours of patient listening and friendship for almost 20 years. Dennis Weibling, for believing in me and giving me such freedom to sow seeds. Paul Lange, whose curiosity and insights are still so vibrant after all these years (not bad for a "plumber"!). Bruce Hosford, for deep friendship, one of the most can-do people I have ever met.

Thanks to Paul Yager, and my friends in the department of bioengineering at the University of Washington School of Medicine, for giving me opportunity. I'm also grateful to my colleagues at Seattle Pacific University: Frank Kline, Rick Eigenbrod, and Bill Rowley, for a spirit of adventure and for tolerance. Don Nielsen, who knew without a doubt that education really was about brain development. Julia Calhoun, who reigns as the premier example of emotional greatness. Alden Jones, amazing as you are, without whom none of my professional life would work.

And my deepest thanks to my beloved wife Kari, who continually reminds me that love is the thing that makes you smile, even when you are tired. You, dear, are one in a million.

about the author

DR. JOHN J. MEDINA is a developmental molecular biologist focused on the genes involved in human brain development and the genetics of psychiatric disorders. He has spent most of his professional life as a private research consultant, working primarily in the biotechnology and pharmaceutical industries on research related to mental health. Medina holds joint affiliate faculty appointments at Seattle Pacific University, where he is the director of the Brain Center for Applied Learning Research, and at the University of Washington School of Medicine, in its Department of Bioengineering.

Medina was the founding director of the Talaris Research Institute, a Seattle-based research center originally focused on how infants encode and process information at the cognitive, cellular, and molecular levels.

In 2004, Medina was appointed to the rank of affiliate scholar at the National Academy of Engineering. He has been named Outstanding Faculty of the Year at the College of Engineering at the University of Washington; the Merrill Dow/Continuing Medical Education National Teacher of the Year; and, twice, the Bioengineering Student Association Teacher of the Year. Medina has been a consultant to the Education Commission of the States and a regular speaker on the relationship between neurology and education.

Medina's books include: The Genetic Inferno; The Clock of Ages; Depression: How it Happens, How it Heals; What You Need to Know About Alzheimer's; The Outer Limits of Life; Uncovering the Mystery of AIDS; and Of Serotonin, Dopamine and Antipsychotic Medications.

Medina has a lifelong fascination in how the mind reacts to and organizes information. As a husband and as a father of two boys, he has an interest in how the brain sciences might influence the way we teach our children. In addition to his research, consulting, and teaching, Medina speaks often to public officials, business and medical professionals, school boards, and nonprofit leaders.

www.johnmedina.com

Biographical Note

Nicole Stellon O'Donnell is the author of two previous collections of poetry, *Steam Laundry* and *You Are No Longer in Trouble*. Her poems have appeared in *Prairie Schooner, Beloit Poetry Journal, Passages North,* and other literary journals. She received both an Individual Artist Award and an Artist Fellowship from the Rasmuson Foundation, as well as a Boochever Fellowship and an Alaska Literary Award from the Alaska Arts and Culture Foundation. Her teaching has been recognized with a Fulbright Distinguished Award in Teaching and a Heinemann Fellowship. She lives and writes in Fairbanks, Alaska.